Aleksey E. Kuznetsov
Editor

GRAPHENE QUANTUM DOTS (GQDS)

Advances in Research and Applications

nova
science publishers
www.novapublishers.com

DOI: https://doi.org/10.52305/FBQU4467

NOTICE TO THE READER

The Publisher has taken reasonable care in the preparation of this book but makes no expressed or implied warranty of any kind and assumes no responsibility for any errors or omissions. No liability is assumed for incidental or consequential damages in connection with or arising out of information contained in this book. The Publisher shall not be liable for any special, consequential, or exemplary damages resulting, in whole or in part, from the readers' use of, or reliance upon, this material. Any parts of this book based on government reports are so indicated and copyright is claimed for those parts to the extent applicable to compilations of such works.

Independent verification should be sought for any data, advice or recommendations contained in this book. In addition, no responsibility is assumed by the Publisher for any injury and/or damage to persons or property arising from any methods, products, instructions, ideas or otherwise contained in this publication.

This publication is designed to provide accurate and authoritative information with regards to the subject matter covered herein. It is sold with the clear understanding that the Publisher is not engaged in rendering legal or any other professional services. If legal or any other expert assistance is required, the services of a competent person should be sought. FROM A DECLARATION OF PARTICIPANTS JOINTLY ADOPTED BY A COMMITTEE OF THE AMERICAN BAR ASSOCIATION AND A COMMITTEE OF PUBLISHERS.

Library of Congress Cataloging-in-Publication Data

ISBN: 979-8-89113-638-0 (hardcover)
ISBN: 979-8-89113-724-0 (e-book)

Published by Nova Science Publishers, Inc. † New York

I dedicate this book to my wife, Camila, and my son, Éomer Evgeniy.

Contents

Preface

Graphene quantum dots (GQDs) are carbon-based nanoparticles with a size generally less than 100 nm. They can be considered as a member of the graphene family, which can be thought of essentially as one-atom thick crystalline structure composed of sp^2-hybridized carbon atoms held together by σ–bonds in a two-dimensional [2D] framework. GQDs consist of one or several layers of graphene sheets and have noticeable chemical and physical stability along with a large surface to mass ratio. Due to the number of exceptional properties such as low toxicity, stable photoluminescence, chemical stability, pronounced quantum confinement effect, etc., GQDs have been applied in various areas, for example, as novel materials in biomedical, optoelectronics, energy related, and environmental applications.

Thus, GQDs have been the subject of significant research interest, and numerous studies have been devoted to them, both theoretical and experimental, from purely academic and practical application points of view. Our book makes its own valuable contribution to this area of knowledge and technology, covering various facets of research and applications of GQDs. Chapter 1 by Jin et al. considers the aspects of synthesis and characterization of graphene quantum dots. Chapter 2 by Jovanović and Dorontić is devoted to the use of GQDs for detection of different analytes, from inorganic ions via biomolecules to pathogenic microorganisms. Chapter 3 by Ramakrishnan and co-authors considers in detail the role of graphene quantum dots in different biological and energy related applications. Chapter 4 by Edet et al. reviews the recent progress in the research of graphene quantum dots for energy applications and storage devices. Chapter 5 by Devi et al. covers the aspects of using graphene quantum dots as promising catalysts for electrocatalytic water splitting. Chapter 6 by Senel and Ibraimi considers uses of GQDs for various biological and biomedical applications. Finally, Chapter 7 by Rani and Srivastava covers different aspects of investigation and applications of GQDs.

We strongly believe that our book will be highly attractive for the broad audience of students and researchers interested in the ever-developing area of the graphene quantum dots, providing them with the useful knowledge of various characteristics of GQDs and peculiarities of their synthesis and characterization. Moreover, our book will provide readers with a review of the history of GQD investigations, the current state-of-the-art of their research and different applications, and perspectives in this field of science and technology.

Aleksey E. Kuznetsov
Editor

Chapter 1

Synthesis and Characterization of Graphene Quantum Dots

Xinjian Jin[1]
Zhi Liu[1]
Shuguang Deng[2,*]
and Gui-Ping Dai[1,†]
[1]Department of Chemical Engineering, School of Chemistry and Chemical Engineering, Nanchang University, Nanchang, China
[2]School for Engineering of Matter, Transport and Energy, Arizona State University, Tempe, AZ, USA

Abstract

As an emerging class of carbon nanomaterials, graphene quantum dots (GQDs) with a size below 10 nm have attracted enormous attention so far due to their advantages of low cost, biocompatibility, and unique photoelectric properties. They have been widely applied in various fields including but not limited to sensors, biomedicine, photoelectrocatalysis, and energy conversion. In this chapter, the methods of synthesis and characterization are briefly summarized, which are critical for designing and understanding of GQDs. Furthermore, the merits of each preparation technique and roles of each characterization method are discussed. At the end of the chapter, current obstacles, and future perspective of GQDs in synthesis and characterization are pointed out.

Keywords: graphene quantum dots, synthesis, characterization

* Corresponding Author's Email: Shuguang.Deng@asu.edu.
† Corresponding Author's Email: nanodai@gmail.com.

In: Graphene Quantum Dots (GQDs)
Editor: Aleksey Kuznetsov
ISBN: 979-8-89113-638-0

1. Introduction

Graphene has been systematically studied by many researchers since Novoselov and Geim separated graphene from graphite with tape in 2004 (Xu et al. 2004). Graphene, a two-dimensional lattice of sp^2 hybridized carbon (see Figure 1), has been widely explored for a variety of applications in diverse fields. Graphene exhibits excellent electrical properties, optical properties, ultra-high thermal conductivity, and good air permeability (X.M. Li et al. 2015). Graphene also has excellent mechanical strength and flexibility due to its strong covalent bonds within the plane (C. Lee et al. 2008). It exhibits special properties such as large specific surface area, single atom thickness, and high electrical conductivity, which has attracted great attention and has broad application prospects in energy storage (Chen et al. 2018), photoelectric devices (Z. Liu et al. 2022), sensors (Faridbod and Sanati 2019) and other fields. However, two-dimensional graphene has no fluorescence due to the existence of a special zero-bandgap structure and difficulty in being dispersed in common solvents (F. Liu et al. 2018). These shortcomings limit its applications in the field of optoelectronic devices and biological imaging.

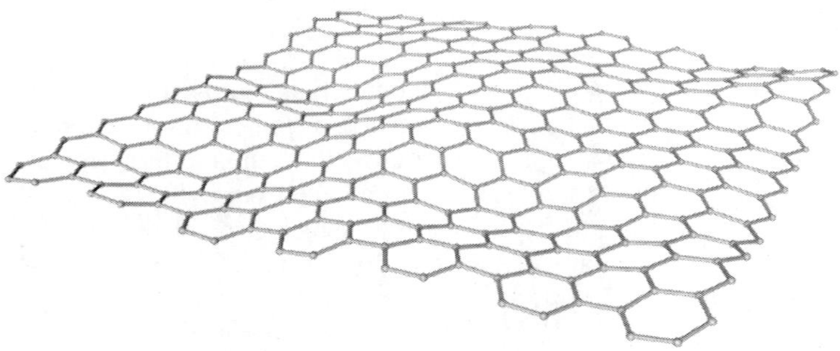

Figure 1. Schematic diagrams of graphene.

Carbon dots (CDs), namely semiconductor nanocrystals, are a class of quasi-zero-dimensional nanomaterials with a particle size below 10 nm, which was discovered by the American Optical Society in 1980 (Resch-Genger et al. 2008). CDs is a broad term used to broadly characterize zero-dimensional carbon nanomaterials with dimensions less than 10 nm. Fluorescent emission in the range from the UV to near-infrared wavelength is obtained by controlling the size and composition of the CDs. The structure of CDs usually consists of a nucleus of sp^2/sp^3 hybridized carbon atoms, and the surface is usually composed of oxygen/nitrogen-based surface functional groups and abundant polymer chains, which make CDs disperse well in water. Various types of CDs could be rapidly synthesized due to diverse synthetic methods (top-down and bottom-up) and a wide range of raw materials (graphite, small molecules, polymers, and even natural materials, etc.) (Zheng et al. 2015). They have the characteristics of large Stokes shift, wide excitation spectrum and narrow emission spectrum (Hu et al. 2022). In addition, CDs possess excellent optical properties and stability against photobleaching. Their electrons and holes are quantum limited due to the small size of CDs, which show many special properties and have been well used in the light emitting diodes (LEDs), bioimaging, photocatalysis, sensing, lasers, and other fields (see Figure 2) (He et al. 2022; Wang et al. 2020; Liu et al. 2019; Xian and Li 2022). Currently, the nomenclature of CDs varies from literature to literature, mainly focusing on CDs, carbon nanoparticles, GQDs, carbon quantum dots, carbon nanodots, polymer dots and carbonized polymer dots, etc. (Tao et al. 2019; Verma, Yadav and Nandi 2019; Xia et al. 2019). Despite the extensive debate on the classification and nomenclature of CDs, CDs are clearly divided into graphene quantum dots (GQDs), carbon quantum dots (CQDs), polymer dots (PDs) and carbonized polymer dots (CPDs) in many research articles and reviews (Zeng et al. 2021). Although there are many classifications for CDs, these classifications are not independent. Both GQDs and CQDs have basic layered structure of graphite. The degree of exfoliation of CQDs determines whether the final product is anisotropic GQDs (several layers thick) or CQDs (multilayer thickness). Rapid aggregation and structural transformation may happen if GQDs are not properly stabilized, and they may eventually transform into amorphous CQDs.

GQDs, as a branch of graphene research, are a new type of carbon fluorescent nanomaterial with a lateral size of less than 100 nm and graphene layer number below 10 layers. GQDs have one or several layers of graphene at the edges and edge-connected chemical groups, whose lateral dimensions are larger than their heights. In particular, it is worth noting that the distinction between GQDs and CQDs is not determined by the presence of a lattice, but by the presence of quantum size effects. Sometimes the existing lattice does not necessarily correspond to the lattice constant of graphite, which may result in polymer nanocrystals.

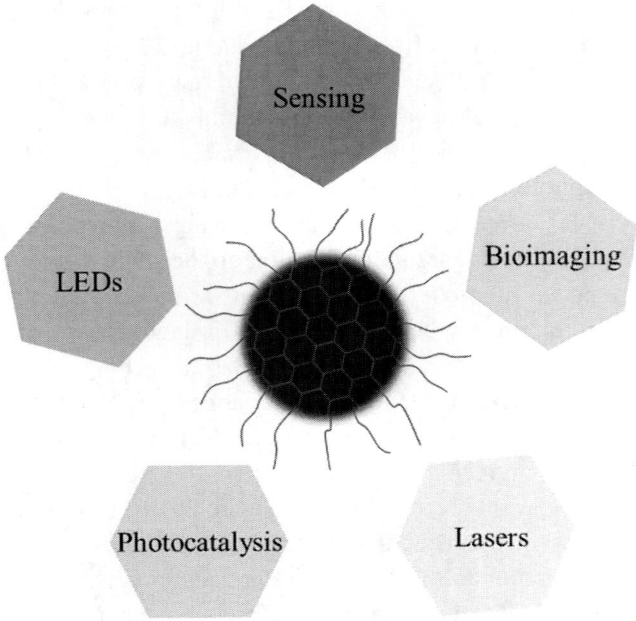

Figure 2. The application fields of CDs.

The diversity of GQDs is mainly due to the differences in carbon sources and synthesis methods. At present, GQDs synthesis can be mainly divided into two approaches — top-down and bottom-up methods (Valappil, Pillai and Alwarappan 2017). Top-down methods are realized by cutting various cheap and readily available carbon materials of large size into small fragments. Bottom-up methods generate GQDs by stepwise polymerization or pyrolysis carbonization with small carbon-containing molecules as carbon sources. It is worth noting that the top-down synthesis strategy is beneficial to the synthesis of graphene dots or carbon nanodots with high crystallinity and complete structures (Wu et al. 2014; Zhao et al. 2008). The bottom-up strategy usually yields GQDs with amorphous carbon cores, abundant doping sites and numerous surface functional groups. The morphological characteristics, properties, and applications of GQDs materials are closely related to their synthesis methods. The development of efficient synthetic methods is beneficial to solving problems in the field of GQDs, such as precise control of GQDs, large-scale preparation, luminescence mechanism, and fine structure of luminescent groups.

GQDs not only have the sheet structure of graphene, but also have quantum confinement effect (Zheng et al. 2021). It is worth noting that although CDs are also a kind of zero-dimensional material and belong to the carbon nanofamily, their morphology and properties are not exactly the same as those of GQDs. The CDs are spherical carbon particles, some of which are composed of completely amorphous sp^3-hybridized carbon, while others are amorphous sp^3-hybridized carbon shells wrapped around sp^2-hybridized crystalline carbon cores (B.Y. Wang et al. 2021). The carbon atoms in GQDs in graphene crystals are all sp^2 hybridized. The electronic bandgap structure of GQDs is largely affected by quantum confinement effects and determined by the surface energy trap presence. GQDs not only have excellent electrical properties, low toxicity, excellent mechanical strength, but also overcome the shortcoming of poor electronic transmission performance of traditional quantum dots and the problem of the zero-band gap of graphene. As a result, GQDs have a great potential application prospect in the field of bioimaging, sensors, catalysis, and photovoltaic devices (Shen et al. 2012). In this chapter, the methods of synthesis and characterization are briefly summarized, which are critical for designing and applying GQDs.

2. Preparation of GQDs

As mentioned above, the preparation methods of GQDs can be roughly divided into two categories according to the selected carbon source materials — top-down methods and bottom-up methods. The top-down methods use large-sized carbon materials such as graphite as raw materials, and cutting, peeling, or decomposing these materials to obtain GQDs by physical and chemical methods. The bottom-up methods use small carbon-containing molecules as carbon sources to prepare GQDs by stepwise polymerization or pyrolysis carbonization.

2.1. Top-Down Methods

The top-down methods are performed by cutting various cheap and readily available carbon materials of large size (graphite, graphene, carbon nanotubes, carbon black, coal, etc.) into small fragments or small-size nanoparticles. Although the top-down methods have advantages of abundant raw materials and simple technique, it is difficult to precisely control the experimental conditions in specific synthesis process. Normally a broad size distribution and relatively low photoluminescence (PL) quantum yield (QY) can be found in the obtained GQDs as a result (Lou et al. 2020). In addition, although dialysis and other methods can help to classify the obtained GQDs products, the classification process is time-consuming and labor-intensive, which limits the wide application of this synthesis strategy to a certain extent. The top-down methods mainly include acid oxidation, electrochemical, microwave assisted method, hydrothermal methods, etc.

2.2. Acid Oxidation Method

Applying strong acid oxidative exfoliation of carbon-based materials is a common method for the preparation of GQDs, and it is also a typical way for their large-scale production (Lin, Song et al. 2015). Generally, concentrated H_2SO_4/HNO_3 is used to oxidize and destroy the C=C double bonds in carbon materials such as graphite, carbon nanotubes, and graphene, and cut these carbon materials into small-sized GQDs. Peng et al. (Peng et al. 2012) prepared GQDs emitting different colors by adjusting the temperature and oxidatively shearing carbon fibers with concentrated H_2SO_4/HNO_3 (see Figure 3). The size of as-prepared GQDs ranged from 1 nm to 4 nm. Dong et al. (Dong, Chen et al. 2012) used CX-72 carbon black as the carbon source, oxidized and refluxed in concentrated nitric acid, and prepared GQDs with green and yellow fluorescence by oxidatively cleaving the carbon black with concentrated nitric acid. Although it is a simple and feasible method to prepare GQDs by oxidative cleavage of strong acid, strong acid is highly corrosive and dangerous to operate. At the same time, the process of subsequent separation and purification of products is complicated and tedious.

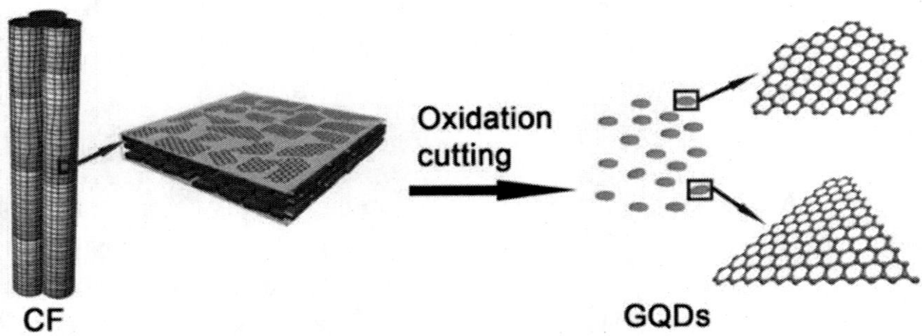

Figure 3. Representation scheme of oxidation cutting of CF into GQDs. Reproduced from https://doi.org/10.1021/nl2038979.

2.3. Electrochemical Method

Figure 4. A facile means to generate fluorescent carbon nanoribbons, nanoparticles, and graphene from graphite electrode. Reproduced from https://doi.org/10.1021/nn900546b.

Electrochemical methods are also commonly used to prepare GQDs (Xie et al. 2020). The process is designed to oxidize or hydroxylate the free valences at the edges and defect areas of the graphite grains of the anode rod with hydroxyl and oxygen radicals generated by electrolytic oxidation of the anode with water, resulting in the decomposition of carbon nanocrystals. As the reaction progresses, the graphite boundary layer is further opened, where anions are inserted into. The graphite anode polarization disappears, and the anode expands and gradually cracks into graphene nanoparticles. Yan et al. (Li et al. 2011) used electrochemical method to fabricate green-luminescent GQDs with uniform size in the range of 3-5 nm. Lu et al. (Lu et al. 2009) successfully prepared fluorescent carbon nanoribbons, nanoparticles, and graphene by electrolysis in aqueous BF_4^- ionic solution with graphite as an electrode (see Figure 4), Xie et al. (Xie et al. 2020) successfully prepared nitrogen-doped GQDs (N-GQDs) using the N-containing 3D nanocarbon hybrids as the carbon and nitrogen sources via an electrochemical method.

Results indicated that the synthesized N-GQDs possessed highly crystallized structure and a high photoluminescence quantum yield of 19.3%. The as-prepared GQDs obtained through electrochemical method are highly stable and remain stable for several months at room temperature without aggregation and color change (Kumar, Sadique and Khan 2021). However, this method has several problems — the pretreatment of raw material graphite takes a long time, the dialysis steps required for the purification of GQDs in the later stage take a long time, and the product yield is not high.

2.4. Microwave-Assisted Method

The microwave synthesis method means irradiating carbon precursors with microwaves to obtain GQDs (Xie et al. 2020). Microwave radiation provides a fast and uniform heating method, which shortens the reaction time and greatly improves the yield and purity of GQDs. The principle of method is that under the action of the electromagnetic field, dielectric material molecules composed of polar and non-polar molecules change from the original random distribution state to an orientation based on the polarity of the electric field. These orientations change continuously based on the alternating electromagnetic frequency, which causes the motion and friction of the molecules to generate heat. Subsequently, the field energy of the alternating electric field is converted into heat energy in the medium, so that the temperature of the medium continues to rise and achieve the purpose of synthesizing GQDs. Kumar et al. (Kumar et al. 2022) fabricated GQDs from Styrofoam waste by a simple microwave-assisted pyrolytic process without acid/base-free (see Figure 5). The result indicated that the as-prepared GQDs were controlled in size and did not have any polar functionalities. Importantly, the structure of GQDs was confirmed to be single layered with an average size of 5.5 ± 1.5 nm. Hoang et al. (Hoang, Pham and Tran 2020) reported a microwave-assisted hydrothermal method for synthesis of GQDs by using reducing agent NH_3 and thermal energy from microwave oven to oxidize raw graphene oxide. The synthesized GQDs were utilized as a hole transport layer in organic solar cells to enhance the cell quantum efficiency.

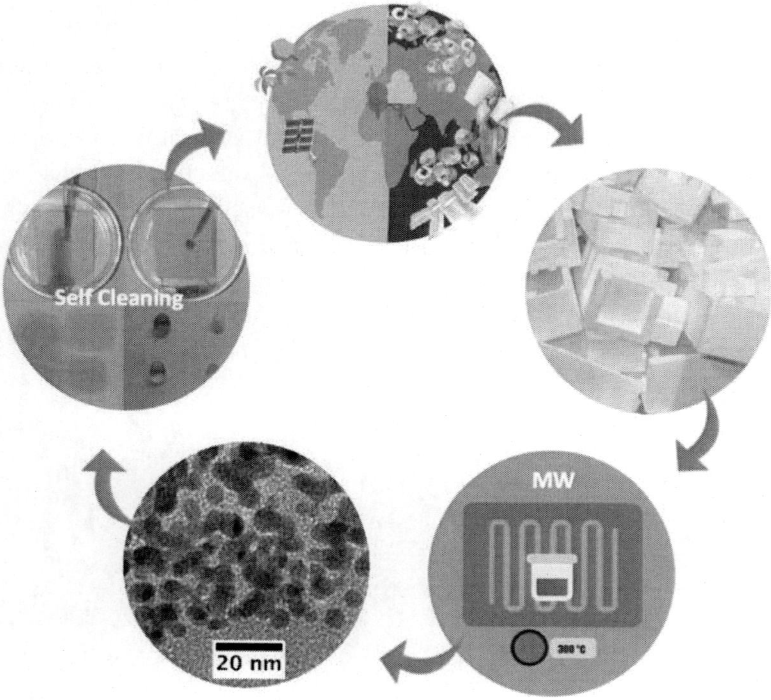

Figure 5. The synthetic of GQDs from Styrofoam waste by a simple microwave-assisted pyrolytic process without acid/base-free Reproduced from https://doi.org/10.1021/acssuschemeng.1c08002.

2.5. Hydrothermal Methods

It is easy to introduce carboxyl groups at the sheet edges and holes of graphene after oxidation by strong acid, and epoxy and carbonyl groups are introduced at the substrate position. These groups form a chemical chain composed of more carbonyl groups and a small amount of epoxy groups on the carbon lattice along the lateral dimension direction. This chain surrounds the surrounding sp^2 clusters and easily causes the C-C bond to break in this region. This chemical chain is easily opened, and the oxygen atom is dissociated in the hydrothermal oxidation environment, and the relatively stable carboxyl group is retained. Finally, GQDs with excellent water solubility are obtained. Pan et al. (Pan et al. 2012) first prepared GQDs with a particle size of about 5-13 nm and blue fluorescence by ultrasonic acid oxidation combined with hydrothermal method. However, the QY was only 7% and fluorescence emission was susceptible to solution pH. Ye et al. (Ye et al. 2013) reported a method of hydrothermal synthesis of GQDs by using anthracite, bituminous coal, and coke as carbon precursors, ultrasonically treated with a mixed solution of concentrated sulfuric acid and concentrated nitric acid (see Figure 6). Although high quantum yield can be obtained by the hydrothermal method, it is based on the raw material graphene and its reduction products, which need to be prepared through a series of chemical reactions and oxidation of a large amount of graphite powder. In addition, a large number of reagents are usually added, and the reduction process is time-consuming.

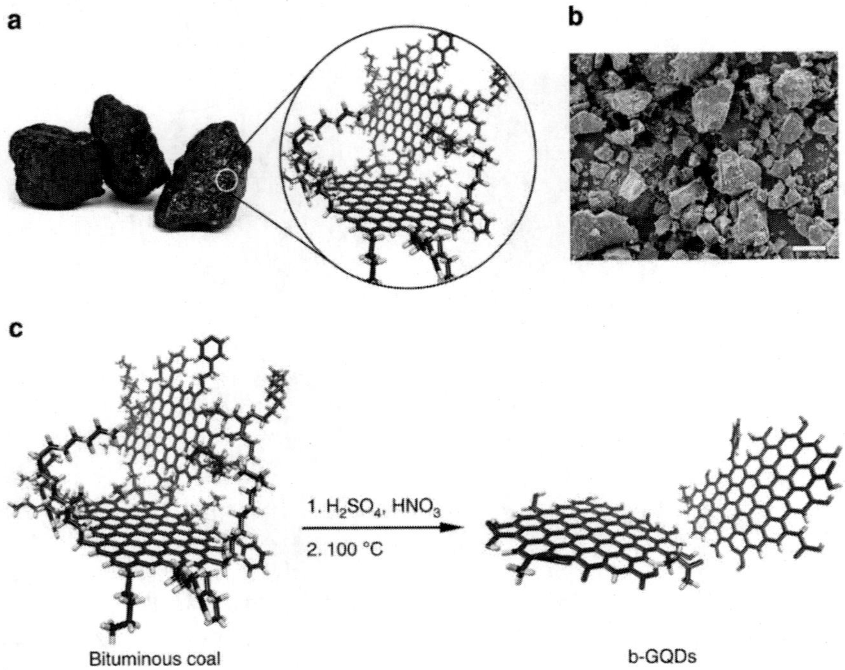

Figure 6. The synthesis of GQDs using coal by hydrothermal synthesis method. Reproduced from 10.1038/ncomms3943.

2.6. Bottom-up Methods

The bottom-up methods exploit small organic molecules as raw materials or precursors to prepare GQDs by pyrolysis, polymerization, etc. The GQDs prepared by these methods commonly contain surface groups, which are conducive to further modification and modification. Recently, the bottom-up methods mainly include pyrolysis, hydrothermal/solvothermal, microwave-assisted method and stepwise organic synthesis method, etc.

2.7. Pyrolysis Method

Some thermodynamically unstable small molecules can be directly carbonized to form GQDs at high temperature, so this approach has been successfully developed as a fast and efficient method for the synthesis of fluorescent GQDs (Lin, Rong et al. 2015). The pyrolysis method uses small organic molecules such as citric acid (CA) and polycyclic aromatic hydrocarbons (PAHs) as precursors.

The preparation of GQDs by using CA as a precursor has been extensively reported. Dong et al. (Dong, Shao et al. 2012) successfully prepared small-sized and larger-sized GQDs from partially and completely carbonized CA, respectively. Hong et al. (Hong et al. 2018) first prepared GQDs by pyrolysis of trisodium citrate, then performed ultrafiltration on the prepared GQDs and finally obtained blue-emitting single-layer GQDs. Zhang et al. (Zhang et al. 2018) synthesized nitrogen and sulfur co-doped GQDs by a facile pyrolysis method using CA and L-cysteine as precursors. Gu et al. (Gu et al. 2020) reported a method for the preparation of highly amidated GQDs by pyrolysis of CA and urea at 250 °C for 0.5 h with infrared assistance.

PAHs can also be used to prepare GQDs. Liu et al. (Liu et al. 2011) prepared nano-sized graphene sheets using hexaphenylbenzene (HBC) as a precursor. The as-prepared GQDs exhibited multicolor fluorescence and layer-to-layer formation by π–π stacking. The morphology of GQDs obtained from the pyrolysis of polycyclic aromatic hydrocarbons is easily affected by the temperature. The GQDs prepared by this method have high fluorescence QY, are well-dispersed and have stable performance, thus they can be stored for a long time (Wang et al. 2019).

2.8. Hydrothermal/Solvothermal Method

Hydrothermal/solvothermal method represents a traditional route to prepare GQDs. GQDs can be synthesized using small molecules in water or organic solvents through condensation, polymerization, carbonization, and passivation processes (Liu et al. 2022; Tam et al. 2014; Xian and Li 2022). The reaction is performed in a Teflon reactor under high temperature and high pressure.

The GQDs prepared by this method have superior purity and follow-up application prospects in many fields.

The hydrothermal method has become the most widely used method for the synthesis of GQDs. Ramachandran et al. (Ramachandranand Karunakaran 2019) synthesized N-GQDs with highly crystallinity and aromaticity by using a single source, polyaniline (PANI), by a simple hydrothermal synthetic strategy. Importantly, the N-doped GQDs (N-GQDs) demonstrated unparalleled electrochemical sensing properties toward nitroaromatics. Tam et al. (Tam et al. 2014) reported a facile bottom-up approach for producing N-GQD carried out by the carbonization of CA through hydrothermal treatment in the presence of ammonia and a fluorescent sensor for determination of Fe^{3+} ions based on these GQDs was reported.

The solvothermal method was expanded on the basis of the hydrothermal method, which means that the original mixture reacts with or non-aqueous solvent in a closed system at a certain temperature and under an autogenic pressure generated by the solution (Dejpasand et al. 2021). The solvothermal method can accurately control the size, shape distribution, and crystallinity of the nanostructures obtained. Qu et al. (Qu et al. 2015b) prepared N-GQDs with tailoring color (blue, green, and yellow) of their emission by simply changing the reaction solvent in the hydrothermal reaction. They believed that larger effective conjugation length combined with a nitrogenous functional group would result in the redshift of emission wavelength. Different sizes of the GQDs with different sizes of conjugated sp^2-domains result in different energy gaps, which determined the corresponding emission colors. Recently, Liu et al. (Liu et al. 2022) employed 1,5-diaminonaphthalene and tartaric acid as the precursors to synthesize single-component white-light-emitting GQDs by a facile one-step solvothermal method (see Figure 7). Importantly, the single-component white-light-emitting GQDs have a PLQY as high as 28.68%. Finally, these GQDs were employed as the luminescence material to fabricate film and white light emitting diodes (WLED).

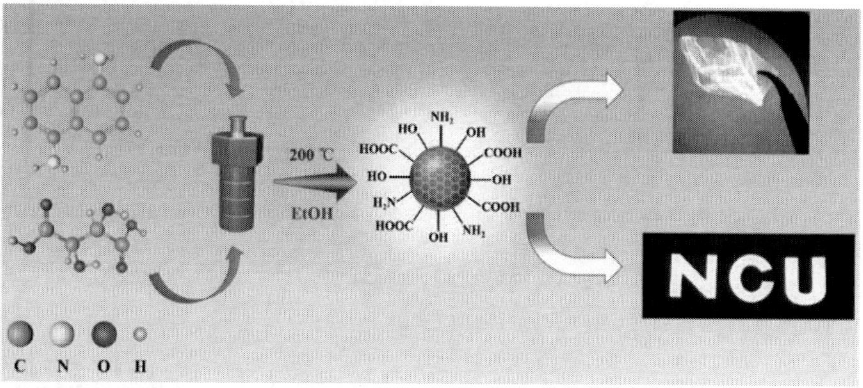

Figure 7. Schematic process of synthesizing W-GQDs by solvothermal treatment. Reproduced from https://doi.org/10.1016/j.optmat.2022.112368.

2.9. Microwave-Assisted Method

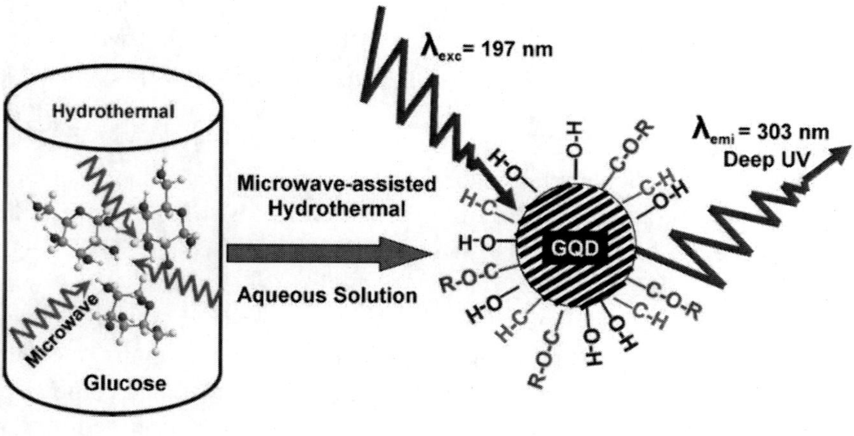

Figure 8. GQDs prepared by a facile microwave-assisted hydrothermal method. Reproduced from https://doi.org/10.1021/nn300760g.

Compared with the hydrothermal synthesis method, the microwave-assisted method can effectively shorten the fusion time of the precursor (Wang et al. 2019). At the same time, this method provides simultaneous heating, which can fabricate GQDs with uniform particle size distribution. Tang et al. (Tang et al. 2012) reported the method of microwave-assisted method at first time (see Figure 8). They found that the emission wavelength of the GQDs was not related to the size of the GQDs, which is different from the conventional carbon quantum dots. The self-passivated layer of GQDs leads to the formation of surface states, which endows the GQDs with superior optical properties. Jeon et al. (Jeon et al. 2016) synthesized N-GQDs and N,S co-doped GQDs using norepinephrine by changing the solvent under microwave irradiation. The resulting N-GQDs and N,S co-doped GQDs showed different photocatalytic and physical properties.

2.10. Stepwise Organic Synthesis Method

The stepwise organic synthesis method is also known as the solution chemistry method (Liu et al. 2022). It is a typical bottom-up GQDs system preparation method, mainly using polycyclic aromatic paths as precursors for controllable synthesis of GQDs. Lee et al. (Lee et al. 2012) used phenyl small molecules as precursors to form covalent bonds with oxygen-containing groups at the edge of graphene and polymerized them to prepare conjugated GQDs with carbon atoms numbers 168, 132, and 170. However, this method is relatively complex. Yan et al. (Yan, Cui and Li 2010) fabricated GQDs with specific structures and sizes by oxidative condensation of aryl groups of polyphenylene dendritic precursors. Lee et al. (Lee et al. 2019) reported on a novel approach for the synthesis of GQDs with the size-controlled single-crystalline via catalytic solution chemistry. In their work, they demonstrated the crucial role of edge states on luminescence from the correlation between the crystalline size and exciton lifetime for the size-controlled single-crystal GQDs. GQDs dots prepared by stepwise organic synthesis method have good water solubility, but with the increase of size, the interaction forces between graphene layers increase, resulting in a decrease in water solubility. Although the size and morphology of GQDs prepared by the method can be precisely controlled, the stepwise organic synthesis method suffers from tedious and complicated procedures.

2.11. Functionalization of GQDs

Although some progress has been made in the preparation and application of GQDs with various emission wavelengths, most GQDs reported showed less than satisfactory optical properties and could not meet the requirements of many current applications (Rabeya et al. 2022). Since most present GQDs show excitation dependence on different factors, the peak value is in the blue-green, and its intensity decreases rapidly with the increase in excitation wavelength. Therefore, it is very important and necessary to functionalize GQDs (see Figure 9).

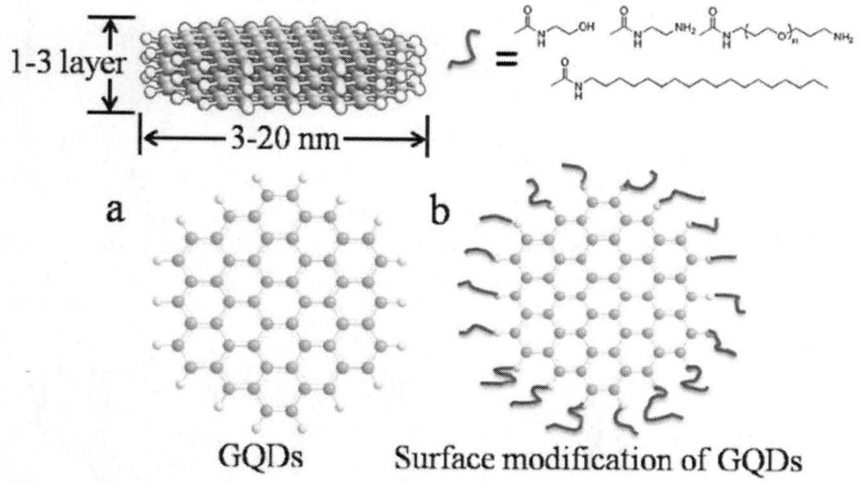

Figure 9. (a) Depiction of GQDs and (b) structural models after functionalization with surface-passivation reagents. Reproduced from https://doi.org/10.1039/C2CC00110A.

2.12. Methods of Functionalization of GQDs via Doping

A large number of studies have reported that doping with heteroatoms plays an important role in regulating and controlling the fluorescence properties of CQDs and adjusting the emission wavelength by modifying the structure and electronic properties of the GQDs (Wang, Zhu and Wang 2019). Since the first report of nitrogen-doped GQDs by Lu et al. in 2012 (Lu et al. 2012), heteroatom doping has become an effective method to regulate and control the properties of GQDs. The carbon atoms in GQDs are replaced by atoms such as boron, nitrogen, sulfur, fluorine, and phosphorus, thereby improving the fluorescence intensity, catalytic performance, and electrical conductivity of GQDs (Qu et al. 2013; Zhang et al. 2018). The co-doping of GQDs by nitrogen and sulfur is the most useful method to construct GQDs with long wavelength emission. Results indicate that sulfur doping can improve the fluorescent intensity, while N-doping can decrease the band gap, ensure the uniformity of the band gap and narrow the emission peak (Sohal, Maity and Basu 2021;

Kaur, Kaur and Sharma 2018). Li et al. (Li et al. 2014) produced monodispersed crystalline sulfur-doped GQDs (S-GQDs) with multiple emission peaks by a hydrothermal method using fructose and sulphuric acid as source materials. The result indicated that sulfur doping can modulate the electronic structure of the GQDs and the effective doping (1% atomic) of sulfur introduced S-related energy levels between π and π^* of C. Recently, Yang et al. (Yang, Wang, Tang et al. 2021) fabricated GQDs with multicolor emission (blue to yellow) via a simple green method. Especially, red emission GQDs with the emission wavelengths extended from green to red are further obtained by sulfur doping. In this work, the authors suggested that the mechanism of red-shift in GQDs after sulfur-doped doping of S can enhance the degree of oxidation and graphitic nitrogen ratio, thus leading to the shift of green to red emission.

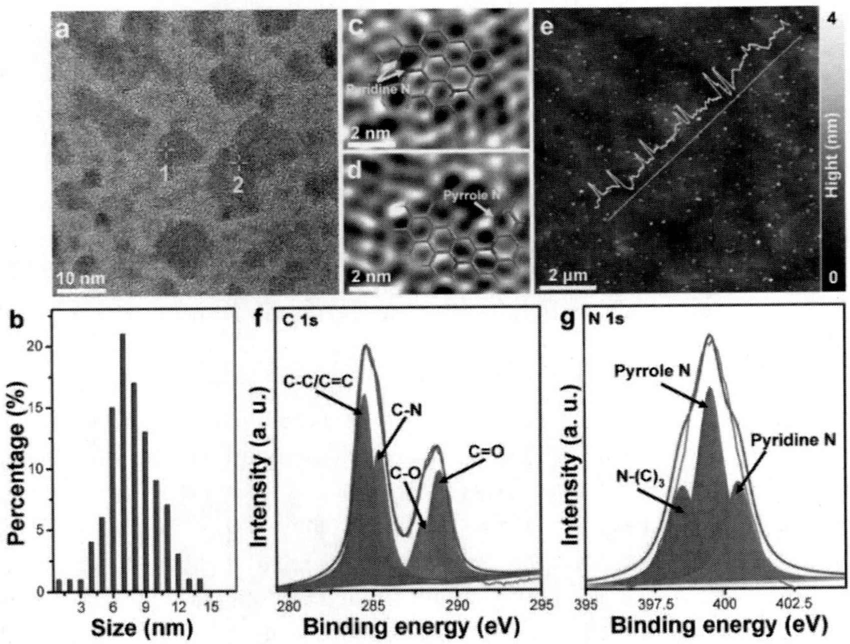

Figure 10. Characterization of N-GQDs. (a) TEM image and (b) lateral size distribution histogram of N-GQDs. (c,d) HAADF-STEM images of single dots labeled in (a). Inset: corresponding chemical structures. (e) Contact-mode AFM image of N-GQDs. Inset: height profile analysis along the line shown in the image. (f) C 1s and (g) N 1s spectra of N-GQDs. Reproduced from https://doi.org/10.1002/smll.202004867.

Researchers have conducted extensive research on nitrogen-doped GQDs as well, since nitrogen atoms are more electronegative than carbon atoms and can change the optical properties of GQDs (Qu et al. 2015a). For example, Yang et al. (Yang, Wang, He, et al. 2021) demonstrated the potential of N-doped graphene quantum dots (N-GQDs) to act as a sonosensitizer. After detailed characterizations, they believed that pyrrole N's and pyridine N's in N-doped GQDs form catalytic centers in sonochemical processing of N-GQDs (see Figure 10).

In addition, since boron and carbon atoms have similar atomic radii, the scientific community is also very interested in the properties of boron-doped GQDs (Wang et al. 2019; Dejpasand et al. 2021). Juang et al. (Juang et al. 2021) synthesized boron-doped GQDs with an average particle size of about 4 nm using 1,3,6-trinitropyrene and boric acid as raw materials by a solvothermal method. The GQDs exhibit good dispersibility in both polar and non-polar solvents due to the existence of nitro groups at the edges of the GQDs. At the same time, they emitted strong green fluorescence and red fluorescence, respectively, when dispersed in *n*-hexane and propylene glycol methyl ether acetate.

2.13. Methods of Functionalization of GQDs via Surface Modification

The properties of GQDs including electronic, optical, and catalytic are affected by various factors such as size, functional groups, and dopants (Yang et al. 2018; Peng et al. 2012). It is the key to obtaining GQDs with excellent properties and ensuring that they can be used effectively by precise regulation of these factors. Surface modification plays a key role in adjusting the optical, electronic, biological, and catalytic properties of GQDs (Zhang et al. 2022). In addition, surface functionalization can change the π-conjugated network and defects on the GQDs surface, affecting adsorption and desorption of gas molecules on it, as well as its hydrophilicity and hydrophobicity. The modification of GQDs is achieved by using organic precursors or intermolecular links containing the required functional groups. Related experimental results indicate that the size of band gap in GQDs can be regulated by adjusting the size of sp^2 domains (Yamato et al. 2019). At the same time, the surface groups (especially oxygen-related functional groups) on the GQDs can also change GQDs bandgap by surface modification (Roeinfard et al. 2020). GQDs emitting at long wavelengths can be obtained by this strategy. Qian et al. (Qian et al. 2013) reported that diamines have a strong ability to adjust the optical properties of GQDs due to their special protonation phenomenon. Tetsuka et al. (Tetsuka et al. 2016) reported that amino functionalization can lead to the redshift of absorption bands. The most attractive properties of GQDs is the tunability of their PL, which is raised by the quantum confinement effect and the surface/edge effect. Wang et al. (Wang et al. 2013) prepared green-emitting polyethylene glycol-functionalized GQDs, which not only have high fluorescence intensity but also can be loaded with drugs through the binding site of polyethylene glycol.

In addition, the surface of GQDs can also be modified by chemical group grafting because most GQDs have oxygen-containing groups at the edges (Sun et al. 2017). Since the introduction of an intermediate n orbitals between the π and π^* orbitals after GQDs are grafted with electron-donating chemical groups, the bandgap of GQDs can be changed, and the stronger the electron-donating ability of the chemical groups, the narrower the resulting GQDs bandgap. Furthermore, the introduction of other chemical groups can also endow GQDs with other functions.

2.14. Methods of Functionalization of GQDs via Compound Functional Components

In addition to doping GQDs with heteroatoms and introducing functional groups, composite functional components can also improve the properties of GQDs which can expand the application of GQDs (Jin et al. 2013). Huo et al. (Huo et al. 2021) synthesized composite TiO_2 hollow nanofibers with GQDs. The photocatalytic performance of the composite was improved, and the ability to degrade methylene blue was significantly enhanced since GQDs exhibit ultrafast electron transport rates while suppressing electron and hole pair recombination. Yin et al. (Yin et al. 2019) designed and fabricated multilayer $NiO@Co_3O_4$ hollow spheres loaded with GQDs. The composite exhibited excellent electrochemical performance in lithium-ion batteries and supercapacitors.

3. Characterization of GQDs

When two-dimensional graphene is converted into zero-dimensional GQDs with a size of only a few nanometers, the distribution of electrons changes significantly, resulting in quantum confinement and edge effects. These unique properties make GQDs different from carbon nanomaterials such as graphene, fullerenes, and carbon nanotubes, and result in them having excellent PL and electrochemical properties.

3.1. Photoluminescent Properties

The photoluminescence (PL) mechanism of GQDs is still under debate. At present, there are two main light-emitting mechanisms mentioned by researchers, namely size effect and surface/boundary effect (Ryu et al. 2015; Zhu et al. 2017). Numerous research results have shown that the luminescence of GQDs is mainly controlled by the core of graphene structure and surface or edge groups in GQDs (Zhu, Shao et al. 2015). Specifically, the core of graphene structure determines the intrinsic bandgap emission of GQDs, while the chemical groups on the surface or edge affect the surface states and thus can change the fluorescence emission of GQDs, which is the surface state luminescence. For GQDs with few surface groups, the band gap of the conjugated π domain is considered to be the fluorescence emission center. Recently, Yan et al. (Yan, Cui and Li 2010) synthesized GQDs with the size of uniformly and tunable via a chemical route. The study found that emission wavelength red-shifted with the increase of size in GQDs. Zhu et al. (Zhu et al. 2014) synthesized GQDs with inconsistent carbon numbers, the result demonstrated that the eigenstates depend on the size of GQDs.

The PL behavior of GQDs is affected by various factors, such as the size and shape of the GQDs, the π-conjugated structural region, the pH of the medium, the type of graphene sheet edges, the degree of oxidation, and surface functionalization type (Al Jahdaly et al. 2021; Jang et al. 2015). Generally, the fluorescence quantum yield of GQDs can indicate the intensity of fluorescence. Studies showed that the reduction of GQDs can not only remove oxygen-containing groups, but also increase the quantum yield. With the removal of oxygen-containing groups, the sp^2 structure region of GQDs increases, and the π-conjugated system also increases. Feng et al. (Feng et al. 2014) enhanced the fluorescence of GQDs by hydrazine hydrate reduction, and the fluorescence intensity of the prepared reduced GQDs was twice that of the unreduced. GQDs are often influenced by particle size, surface atoms, edge shape, and electronic structure, resulting in excitation wavelength-dependent PL behavior (Tian et al. 2018). The emission wavelength of GQDs is related to their bandgap, the released energy when the electron returns to the ground state increases with expanding of bandgap (Zhu, Song, et al. 2015). GQDs have a quantum confinement effect, and the bandgap increases with decreasing size. GQDs can emit light of different wavelengths by adjusting the bandgap, among which mostly blue and green emissions are present. At present, there are many studies on the effect of functionalization and modification on the surface of GQDs on luminescence behavior (Li et al. 2018; Liu et al. 2020). For example, Naksen et al. (Naksen et al. 2022) reported N-GQDs as a fluorescent sensor for Cd(II) determination. N-GQDs were synthesized from CA and ethylenediamine via the hydrothermal method (see Figure 11).

In addition to the above-presented results, the fluorescence emission of GQDs also was shown to exhibit other characteristics such as luminescence, excitation wavelength dependence and pH-dependence (Routh et al. 2013; Xie et al. 2020). Up-conversion luminescence refers to the emission of short-wavelength light with higher energy when the luminescent material is excited by long-wavelength light with lower energy. Compared with organic small molecule fluorophores, GQDs have higher up-conversion luminescence efficiency. Considering the enormous biomedical application of GQDs, the long wavelength-excited emission is highly desired because that will effectively avoid of bioluminescence (Qu et al. 2015b). Therefore, GQDs with converted luminescence properties are ideal fluorescent materials for in vivo imaging. Most GQDs exhibit excitation wavelength dependence with a change from blue to red fluorescence under different excitation wavelengths. pH-dependence is another property of the fluorescence emission of GQDs. The changes in pH also impact the PL behavior of GQDs when the existence of functional groups can be protonated or deprotonated in GQDs (Valappil, Pillai and Alwarappan 2017). The pH-dependent properties of the fluorescence of GQDs are usually not related to their size, but to the structure of their surface groups, and GQDs containing different groups show different pH-dependence.

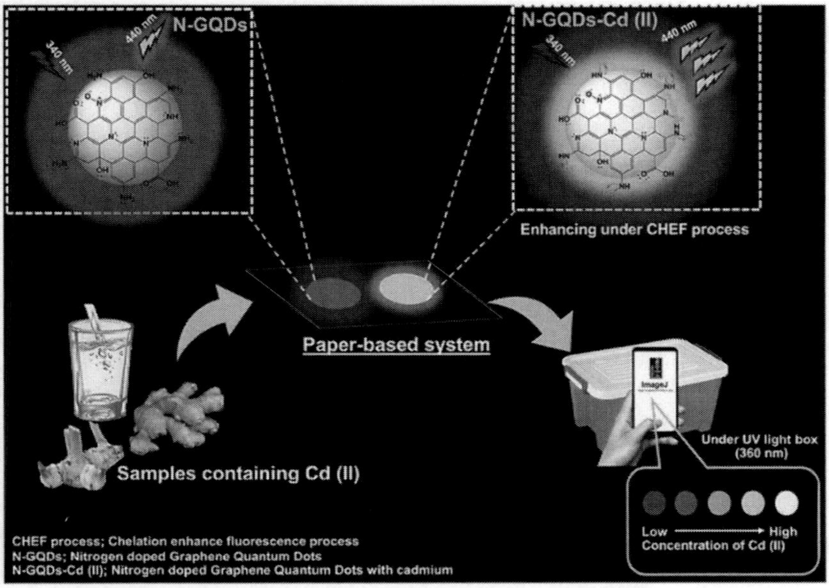

Figure 11. Turn-on fluorescence of N-GQDs based on the CHEF mechanism. Reproduced from https://doi.org/10.1016/j.talanta.2022.123305.

3.2. Electrochemical Properties

GQDs exhibit excellent performance in the field of electrochemistry because have a large specific surface area, abundant binding sites and good electrical conductivity (Kumar et al. 2020). On the other hand, GQDs have large surface-to-volume ratio and appropriate charge transferability. Electrochemical biosensors based on GQDs have high performance (Tang et al. 2019). In addition, they also have electroluminescent properties. Li et al. (Li et al. 2012) prepared GQDs with green fluorescence and showed a strong electroluminescence signal, but no signal was observed with the same concentration of graphene oxide.

3.3. Dispersion Properties

GQDs can be easily dispersed in various solvents and thus applied in various fields (Dan et al. 2015). This property can be attributed to the following factors: there are negatively charged oxygen functional groups in GQDs, which help GQDs to avoid the aggregation (Zhang et al. 2012); some functional groups increase the polarity of GQDs, which is beneficial to the dispersion of GQDs in water or other polar solvents (Zhu et al. 2011); the formation of hydrogen bonds is possible between GQDs and solvent molecules (Mei et al. 2019). The smaller the size of GQDs, the smaller the contribution of their graphene structure. Dispersibility in solvents can be improved by changing the graphene edge structure.

4. Current Challenges and Future Prospects

Due to their unique and superior features GQDs have great potential in many future technological fields. The preparation and characterization of GQDs have become one of the attractions in the field. Researchers have used a variety of preparation methods to prepare GQDs with different sizes and edge morphologies and to study their properties. However, the large-scale preparation and practical application of GQDs are facing many problems, and there are still many practical problems that need to be solved in order to make them truly enter the application field.

References

Al Jahdaly, B. A., M. F. Elsadek, B. M. Ahmed, M. F. Farahat, M. M. Taher, and A. M. Khalil. 2021. "Outstanding Graphene Quantum Dots from Carbon Source for Biomedical and Corrosion Inhibition Applications: A Review." *Sustainability* 13 (4). https://doi.org/10.3390/su13042127.

Chen, K. N., Q. R. Wang, Z. Q. Niu, and J. Chen. 2018. "Graphene-based materials for flexible energy storage devices." *Journal of Energy Chemistry* 27 (1): 12-24. https://doi.org/10.1016/j.jechem.2017.08.015.

Dan, Qu, Zheng Min, Li Jing, Xie Zhigang, and Sun Zaicheng. 2015. "Tailoring color emissions from N-doped graphene quantum dots for bioimaging applications." *Light:Science & Applications* 4 (4): 137.

Dejpasand, M. T., S. Sharifi, E. Saievar-Iranizad, A. Yazdani, and K. Rahimi. 2021. "Boron- and nitrogen-doped graphene quantum dots with enhanced supercapacitance." *Journal of Energy Storage* 42. https://doi.org/10.1016/j.est.2021.103103.

Dong, Y. Q., C. Q. Chen, X. T. Zheng, L. L. Gao, Z. M. Cui, H. B. Yang, C. X. Guo, Y. W. Chi, and C. M. Li. 2012. "One-step and high yield simultaneous preparation of single- and multi-layer graphene quantum dots from CX-72 carbon black." *Journal of Materials Chemistry* 22 (18): 8764-8766. https://doi.org/10.1039/c2jm30658a.

Dong, Y. Q., J. W. Shao, C. Q. Chen, H. Li, R. X. Wang, Y. W. Chi, X. M. Lin, and G. N. Chen. 2012. "Blue luminescent graphene quantum dots and graphene oxide prepared by tuning the carbonization degree of citric acid." *Carbon* 50 (12): 4738-4743. https://doi.org/10.1016/j.carbon.2012.06.002.

Faridbod, Farnoush, and Afsaneh L. Sanati. 2019. "Graphene Quantum Dots in Electrochemical Sensors/Biosensors." *Current Analytical Chemistry* 15 (2): 103-123. https://doi.org/10.2174/1573411014666180319145506.

Feng, Y. Q., J. P. Zhao, X. B. Yan, F. L. Tang, and Q. J. Xue. 2014. "Enhancement in the fluorescence of graphene quantum dots by hydrazine hydrate reduction." *Carbon* 66: 334-339. https://doi.org/10.1016/j.carbon.2013.09.008.

Gu, S. Y., C. T. Hsieh, B. C. Mallick, and Y. G. Ashraf. 2020. "Infrared-assisted synthesis of highly amidized graphene quantum dots as metal-free electrochemical catalysts." *Electrochimica Acta* 360. https://doi.org/10.1016/j.electacta.2020.137009.

He, C., X. F. Lin, Y. Q. Mei, Y. Luo, M. Yang, Y. Kuang, X. Q. Yi, W. J. Zeng, Q. T. Huang, and B. Zhong. 2022. "Recent Advances in Carbon Dots for In Vitro/Vivo Fluorescent Bioimaging: A Mini-Review." *Frontiers in Chemistry* 10. https://doi.org/10.3389/fchem.2022.905475.

Hoang, Thi Thu, Hoai Phuong Pham, and Quang Trung Tran. 2020. "A Facile Microwave-Assisted Hydrothermal Synthesis of Graphene Quantum Dots for Organic Solar Cell Efficiency Improvement." *Journal of Nanomaterials* 2020: 1-8. https://doi.org/10.1155/2020/3207909.

Hong, G. L., H. L. Zhao, H. H. Deng, H. J. Yang, H. P. Peng, Y. H. Liu, and W. Chen. 2018. "Fabrication of ultra-small monolayer graphene quantum dots by pyrolysis of trisodium citrate for fluorescent cell imaging." *International Journal of Nanomedicine* 13: 4807-4815. https://doi.org/10.2147/ijn.S168570.

Hu, J. Y., Y. F. Guo, X. Geng, J. L. Wang, S. Li, Y. Q. Sun, L. B. Qu, and Z. H. Li. 2022. "Tuning asymmetric electronic structure endows carbon dots with unexpected huge stokes shift for high contrast in vivo imaging." *Chemical Engineering Journal* 446. https://doi.org/10.1016/j.cej.2022.136928.

Huo, P. P., P. Zhao, X. B. Shi, Z. Y. Zhou, and B. Liu. 2021. "Enhanced photocatalytic performance of electrospun hollow titanium dioxide nanofibers decorated with graphene quantum dots." *Journal of Materials Science* 56 (3): 2138-2149. https://doi.org/10.1007/s10853-020-05352-5.

Jang, M. H., H. D. Ha, E. S. Lee, F. Liu, Y. H. Kim, T. S. Seo, and Y. H. Cho. 2015. "Is the Chain of Oxidation and Reduction Process Reversible in Luminescent Graphene Quantum Dots?" *Small* 11 (31): 3773-3781. https://doi.org/10.1002/smll.201500206.

Jeon, S. J., T. W. Kang, J. M. Ju, M. J. Kim, J. H. Park, F. Raza, J. Han, H. R. Lee, and J. H. Kim. 2016. "Modulating the Photocatalytic Activity of Graphene Quantum Dots via Atomic Tailoring for Highly Enhanced Photocatalysis under Visible Light." *Advanced Functional Materials* 26 (45): 8211-8219. https://doi.org/10.1002/adfm.201603803.

Jin, S. H., D. H. Kim, G. H. Jun, S. H. Hong, and S. Jeon. 2013. "Tuning the Photoluminescence of Graphene Quantum Dots through the Charge Transfer Effect of Functional Groups." *Acs Nano* 7 (2): 1239-1245. https://doi.org/10.1021/nn304675g.

Juang, R. S., C. T. Hsieh, C. P. Kao, Y. A. Gandomi, C. C. Fu, S. H. Liu, and S. Y. Gu. 2021. "Highly fluorescent green and red emissions from boron-doped graphene quantum dots under blue light illumination." *Carbon* 176: 61-70. https://doi.org/10.1016/j.carbon.2021.01.019.

Kaur, M., M. Kaur, and V. K. Sharma. 2018. "Nitrogen-doped graphene and graphene quantum dots: A review onsynthesis and applications in energy, sensors and environment." *Advances in Colloid and Interface Science* 259: 44-64. https://doi.org/10.1016/j.cis.2018.07.001.

Kumar, Dheeraj, Sweety Rani, Bhanu Nandan, and Rajiv K. Srivastava. 2022. "Nonpolar Graphene Quantum Dot-Based Hydrophobic Coating from Microwave-Assisted Treatment of Styrofoam Waste." *ACS Sustainable Chemistry & Engineering* 10 (2): 1070-1077. https://doi.org/10.1021/acssuschemeng.1c08002.

Kumar, N., M. A. Sadique, and R. Khan. 2021. "Electrochemical exfoliation of graphene quantum dots from waste dry cell battery for biosensor applications." *Materials Letters* 305. https://doi.org/10.1016/j.matlet.2021.130829.

Kumar, Y. R., K. Deshmukh, K. K. Sadasivuni, and S. K. K. Pasha. 2020. "Graphene quantum dot based materials for sensing, bio-imaging and energy storage applications: a review." *Rsc Advances* 10 (40): 23861-23898. https://doi.org/10.1039/d0ra03938a.

Lee, C., X. D. Wei, J. W. Kysar, and J. Hone. 2008. "Measurement of the elastic properties and intrinsic strength of monolayer graphene." *Science* 321 (5887): 385-388. https://doi.org/10.1126/science.1157996.

Lee, J., K. Kim, W. I. Park, B. H. Kim, J. H. Park, T. H. Kim, S. Bong, C. H. Kim, G. Chae, M. Jun, Y. Hwang, Y. S. Jung, and S. Jeon. 2012. "Uniform Graphene Quantum Dots Patterned from Self-Assembled Silica Nanodots." *Nano Letters* 12 (12): 6078-6083. https://doi.org/10.1021/nl302520m.

Lee, S. H., D. Y. Kim, J. Lee, S. B. Lee, H. Han, Y. Y. Kim, S. C. Mun, S. H. Im, T. H. Kim, and O. O. Park. 2019. "Synthesis of Single-Crystalline Hexagonal Graphene Quantum Dots from Solution Chemistry." *Nano Letters* 19 (8): 5437-5442. https://doi.org/10.1021/acs.nanolett.9b01940.

Li, K. H., X. N. Zhao, G. Wei, and Z. Q. Su. 2018. "Recent Advances in the Cancer Bioimaging with Graphene Quantum Dots." *Current Medicinal Chemistry* 25 (25): 2876-2893. https://doi.org/10.2174/0929867324666170223154145.

Li, L. L., J. Ji, R. Fei, C. Z. Wang, Q. Lu, J. R. Zhang, L. P. Jiang, and J. J. Zhu. 2012. "A Facile Microwave Avenue to Electrochemiluminescent Two-Color Graphene Quantum Dots." *Advanced Functional Materials* 22 (14): 2971-2979. https://doi.org/10.1002/adfm.201200166.

Li, X., S. P. Lau, L. Tang, R. Ji, and P. Yang. 2014. "Sulphur doping: a facile approach to tune the electronic structure and optical properties of graphene quantum dots." *Nanoscale* 6 (10): 5323-8. https://doi.org/10.1039/c4nr00693c.

Li, X. M., M. C. Rui, J. Z. Song, Z. H. Shen, and H. B. Zeng. 2015. "Carbon and Graphene Quantum Dots for Optoelectronic and Energy Devices: A Review." *Advanced Functional Materials* 25 (31): 4929-4947. https://doi.org/10.1002/adfm.201501250.

Li, Y., Y. Hu, Y. Zhao, G. Q. Shi, L. E. Deng, Y. B. Hou, and L. T. Qu. 2011. "An Electrochemical Avenue to Green-Luminescent Graphene Quantum Dots as Potential Electron-Acceptors for Photovoltaics." *Advanced Materials* 23 (6): 776-+. https://doi.org/10.1002/adma.201003819.

Lin, L. P., M. C. Rong, S. S. Lu, X. H. Song, Y. X. Zhong, J. W. Yan, Y. R. Wang, and X. Chen. 2015. "A facile synthesis of highly luminescent nitrogen-doped graphene quantum dots for the detection of 2,4,6-trinitrophenol in aqueous solution." *Nanoscale* 7 (5): 1872-1878. https://doi.org/10.1039/c4nr06365a.

Lin, L. P., X. H. Song, Y. Y. Chen, M. C. Rong, Y. R. Wang, L. Zhao, T. T. Zhao, and X. Chen. 2015. "Europium-decorated graphene quantum dots as a fluorescent probe for label-free, rapid and sensitive detection of Cu2+ and L-cysteine." *Analytica Chimica Acta* 891: 261-268. https://doi.org/10.1016/j.aca.2015.08.011.

Liu, F., F. S. Niu, T. Chen, J. R. Han, Z. Liu, W. R. Yang, Y. H. Xu, and J. Q. Liu. 2018. "One-step electrochemical strategy for in-situ synthesis of S, N-codoped graphene as metal-free catalyst for oxygen reduction reaction." *Carbon* 134: 316-325. https://doi.org/10.1016/j.carbon.2018.04.007.

Liu, M. L., B. B. Chen, C. M. Li, and C. Z. Huang. 2019. "Carbon dots: synthesis, formation mechanism, fluorescence origin and sensing applications." *Green Chemistry* 21 (3): 449-471. https://doi.org/10.1039/c8gc02736f.

Liu, Q. W., J. H. Sun, K. Gao, N. Chen, X. T. Sun, D. Ti, C. C. Bai, R. R. Cui, and L. T. Qu. 2020. "Graphene quantum dots for energy storage and conversion: from fabrication to applications." *Materials Chemistry Frontiers* 4 (2): 421-436. https://doi.org/10.1039/c9qm00553f.

Liu, R. L., D. Q. Wu, X. L. Feng, and K. Mullen. 2011. "Bottom-Up Fabrication of Photoluminescent Graphene Quantum Dots with Uniform Morphology." *Journal of the American Chemical Society* 133 (39): 15221-15223. https://doi.org/10.1021/ja204953k.

Liu, Zhi, Long Tan, Pei-Pei Hou, Xin-Jian Jin, Meng-Chao Li, Qun-Yi Zhou, Peng Liao, Zheling Zeng, Shuguang Deng, and Gui-Ping Dai. 2022. "One-pot synthesis of single-component graphene quantum dots for stable and bright white luminescence films as a phosphor." *Optical Materials* 127. https://doi.org/10.1016/j.optmat.2022.112368.

Lou, Y., J. Y. Ji, A. M. Qin, L. Liao, Z. Y. Li, S. P. Chen, K. Y. Zhang, and J. Ou. 2020. "Cane Molasses Graphene Quantum Dots Passivated by PEG Functionalization for Detection of Metal Ions." *Acs Omega* 5 (12): 6763-6772. https://doi.org/10.1021/acsomega.0c00098.

Lu, J., J. X. Yang, J. Z. Wang, A. L. Lim, S. Wang, and K. P. Loh. 2009. "One-Pot Synthesis of Fluorescent Carbon Nanoribbons, Nanoparticles, and Graphene by the Exfoliation of Graphite in Ionic Liquids." *Acs Nano* 3 (8): 2367-2375. https://doi.org/10.1021/nn900546b.

Lu, J., P. S. E. Yeo, C. K. Gan, P. Wu, and K. P. Loh. 2011. "Transforming C_{60} molecules into graphene quantum dots." *Nature Nanotechnology* 6 (4): 247-252. https://doi.org/10.1038/nnano.2011.30.

Mei, S. L., X. Wei, Z. Hu, C. Wei, D. L. Su, D. Yang, G. L. Zhang, W. L. Zhang, and R. Q. Guo. 2019. "Amphipathic carbon dots with solvent-dependent optical properties and sensing application." *Optical Materials* 89: 224-230. https://doi.org/10.1016/j.optmat.2019.01.021.

Naksen, P., S. Boonruang, N. Yuenyong, H. L. Lee, P. Ramachandran, W. Anutrasakda, M. Amatatongchai, S. Pencharee, and P. Jarujamrus. 2022. "Sensitive detection of trace level Cd (II) triggered by chelation enhanced fluorescence (CHEF) "turn on": Nitrogen-doped graphene quantum dots (N-GQDs) as fluorometric paper-based sensor." *Talanta* 242. https://doi.org/10.1016/j.talanta.2022.123305.

Pan, D. Y., L. Guo, J. C. Zhang, C. Xi, Q. Xue, H. Huang, J. H. Li, Z. W. Zhang, W. J. Yu, Z. W. Chen, Z. Li, and M. H. Wu. 2012. "Cutting sp(2) clusters in graphene sheets into colloidal graphene quantum dots with strong green fluorescence." *Journal of Materials Chemistry* 22 (8): 3314-3318. https://doi.org/10.1039/c2jm16005f.

Peng, J., W. Gao, B. K. Gupta, Z. Liu, R. Romero-Aburto, L. H. Ge, L. Song, L. B. Alemany, X. B. Zhan, G. H. Gao, S. A. Vithayathil, B. A. Kaipparettu, A. A. Marti, T. Hayashi, J. J. Zhu, and P. M. Ajayan. 2012. "Graphene Quantum Dots Derived from Carbon Fibers." *Nano Letters* 12 (2): 844-849. https://doi.org/10.1021/nl2038979.

Qian, Z. S., J. J. Ma, X. Y. Shan, L. X. Shao, J. Zhou, J. R. Chen, and H. Feng. 2013. "Surface functionalization of graphene quantum dots with small organic molecules from photoluminescence modulation to bioimaging applications: an experimental and theoretical investigation." *Rsc Advances* 3 (34): 14571-14579. https://doi.org/10.1039/c3ra42066c.

Qu, D., M. Zheng, P. Du, Y. Zhou, L. G. Zhang, D. Li, H. Q. Tan, Z. Zhao, Z. G. Xie, and Z. C. Sun. 2013. "Highly luminescent S, N co-doped graphene quantum dots with broad visible absorption bands for visible light photocatalysts." *Nanoscale* 5 (24): 12272-12277. https://doi.org/10.1039/c3nr04402e.

Qu, D., M. Zheng, J. Li, Z. G. Xie, and Z. C. Sun. 2015a. "Tailoring color emissions from N-doped graphene quantum dots for bioimaging applications." *Light-Science & Applications* 4. https://doi.org/10.1038/lsa.2015.137.

Qu, Dan, Min Zheng, Jing Li, Zhigang Xie, and Zaicheng Sun. 2015b. "Tailoring color emissions from N-doped graphene quantum dots for bioimaging applications." *Light: Science & Applications* 4 (12): e364-e364. https://doi.org/10.1038/lsa.2015.137.

Rabeya, R., S. Mahalingam, A. Manap, M. Satgunam, M. Akhtaruzzaman, and C. H. Chia. 2022. "Structural defects in graphene quantum dots: A review." *International Journal of Quantum Chemistry* 122 (12). https://doi.org/10.1002/qua.26900.

Ramachandran, Aswathi, Arya Nair J S, and Sandhya Karunakaran Yesodha. 2019. "Polyaniline-Derived Nitrogen-Doped Graphene Quantum Dots for the Ultratrace Level Electrochemical Detection of Trinitrophenol and the Effective Differentiation of Nitroaromatics: Structure Matters." *ACS Sustainable Chemistry & Engineering* 7 (7): 6732-6743. https://doi.org/10.1021/acssuschemeng.8b05996.

Resch-Genger, U., M. Grabolle, S. Cavaliere-Jaricot, R. Nitschke, and T. Nann. 2008. "Quantum dots versus organic dyes as fluorescent labels." *Nature Methods* 5 (9): 763-775. https://doi.org/10.1038/nmeth.1248.

Roeinfard, M., M. Zahedifar, M. Darroudi, A. K. Zak, and E. Sadeghi. 2020. "Preparation and characterization of selenium-decorated graphene quantum dots with high afterglow for application in photodynamic therapy." *Luminescence* 35 (6): 891-896. https://doi.org/10.1002/bio.3798.

Routh, P., S. Das, A. Shit, P. Bairi, P. Das, and A. K. Nandi. 2013. "Graphene Quantum Dots from a Facile Sono-Fenton Reaction and Its Hybrid with a Polythiophene Graft Copolymer toward Photovoltaic Application." *Acs Applied Materials & Interfaces* 5 (23): 12672-12680. https://doi.org/10.1021/am4040174.

Ryu, J., E. Lee, K. Lee, and J. Jang. 2015. "A graphene quantum dots based fluorescent sensor for anthrax biomarker detection and its size dependence." *Journal of Materials Chemistry B* 3 (24): 4865-4870. https://doi.org/10.1039/c5tb00585j.

Shen, J., Y. Zhu, X. Yang, and C. Li. 2012. "Graphene quantum dots: emergent nanolights for bioimaging, sensors, catalysis and photovoltaic devices." *Chem Commun (Camb)* 48 (31): 3686-99. https://doi.org/10.1039/c2cc00110a.

Sohal, N., B. Maity, and S. Basu. 2021. "Recent advances in heteroatom-doped graphene quantum dots for sensing applications." *Rsc Advances* 11 (41): 25586-25615. https://doi.org/10.1039/d1ra04248c.

Sun, L., S. S. Li, W. Ding, Y. W. Yao, X. Y. Yang, and C. Yao. 2017. "Fluorescence detection of cholesterol using a nitrogen-doped graphene quantum dot/chromium picolinate complex-based sensor." *Journal of Materials Chemistry B* 5 (45): 9006-9014. https://doi.org/10.1039/c7tb02037f.

Tam, Tran Van, Nguyen Bao Trung, Hye Ryeon Kim, Jin Suk Chung, and Won Mook Choi. 2014. "One-pot synthesis of N-doped graphene quantum dots as a fluorescent sensing platform for Fe^{3+} ions detection." *Sensors and Actuators B: Chemical* 202: 568-573. https://doi.org/10.1016/j.snb.2014.05.045.

Tang, J., R. Huang, S. B. Zheng, S. X. Jiang, H. Yu, Z. R. Li, and J. F. Wang. 2019. "A sensitive and selective electrochemical sensor based on graphene quantum dots/gold nanoparticles nanocomposite modified electrode for the determination of luteolin in peanut hulls." *Microchemical Journal* 145: 899-907. https://doi.org/10.1016/j.microc.2018.12.006.

Tang, L. B., R. B. Ji, X. K. Cao, J. Y. Lin, H. X. Jiang, X. M. Li, K. S. Teng, C. M. Luk, S. J. Zeng, J. H. Hao, and S. P. Lau. 2012. "Deep Ultraviolet Photoluminescence of Water-Soluble Self-Passivated Graphene Quantum Dots." *Acs Nano* 6 (6): 5102-5110. https://doi.org/10.1021/nn300760g.

Tao, S. Y., T. L. Feng, C. Y. Zheng, S. J. Zhu, and B. Yang. 2019. "Carbonized Polymer Dots: A Brand New Perspective to Recognize Luminescent Carbon-Based Nanomaterials." *Journal of Physical Chemistry Letters* 10 (17): 5182-5188. https://doi.org/10.1021/acs.jpclett.9b01384.

Tetsuka, Hiroyuki, Akihiro Nagoya, Takanori Fukusumi, and Takayuki Matsui. 2016. "Graphene Quantum Dots: Molecularly Designed, Nitrogen-Functionalized Graphene Quantum Dots for Optoelectronic Devices (Adv. Mater. 23/2016)." *Advanced materials (Deerfield Beach, Fla.)* 28 (23): 4755. https://doi.org/10.1002/adma.201670162.

Tian, P., L. Tang, K. S. Teng, and S. P. Lau. 2018. "Graphene quantum dots from chemistry to applications." *Materials Today Chemistry* 10: 221-258. https://doi.org/10.1016/j.mtchem.2018.09.007.

Valappil, M. O., V. K. Pillai, and S. Alwarappan. 2017. "Spotlighting graphene quantum dots and beyond: Synthesis, properties and sensing applications." *Applied Materials Today* 9: 350-371. https://doi.org/10.1016/j.apmt.2017.09.002.

Verma, N. C., A. Yadav, and C. K. Nandi. 2019. "Paving the path to the future of carbogenic nanodots." *Nature Communications* 10. https://doi.org/10.1038/s41467-019-10394-8.

Wang, B. Y., J. K. Yu, L. Z. Sui, S. J. Zhu, Z. Y. Tang, B. Yang, and S. Y. Lu. 2021. "Rational Design of Multi-Color-Emissive Carbon Dots in a Single Reaction System by Hydrothermal." *Advanced Science* 8 (1). https://doi.org/10.1002/advs.202001453.

Wang, J. Y., Y. H. Zhu, and L. Wang. 2019. "Synthesis and Applications of Red-Emissive Carbon Dots." *Chemical Record* 19 (10): 2083-2094. https://doi.org/10.1002/tcr.201800172.

Wang, X. Y., X. L. Yao, A. Narita, and K. Mullen. 2019. "Heteroatom-Doped Nanographenes with Structural Precision." *Accounts of Chemical Research* 52 (9): 2491-2505. https://doi.org/10.1021/acs.accounts.9b00322.

Wang, Z. F., Y. Liu, S. J. Zhen, X. X. Li, W. G. Zhang, X. Sun, B. Y. Xu, X. Wang, Z. H. Gao, and X. G. Meng. 2020. "Gram-Scale Synthesis of 41% Efficient Single-Component White-Light-Emissive Carbonized Polymer Dots with Hybrid Fluorescence/Phosphorescence for White Light-Emitting Diodes." *Advanced Science* 7 (4). https://doi.org/10.1002/advs.201902688.

Wang, Z. H., J. F. Xia, C. F. Zhou, B. Via, Y. Z. Xia, F. F. Zhang, Y. H. Li, L. H. Xia, and J. Tang. 2013. "Synthesis of strongly green-photoluminescent graphene quantum dots for drug carrier." *Colloids and Surfaces B-Biointerfaces* 112: 192-196. https://doi.org/10.1016/j.colsurfb.2013.07.025.

Wu, M. B., Y. Wang, W. T. Wu, C. Hu, X. N. Wang, J. T. Zheng, Z. T. Li, B. Jiang, and J. S. Qiu. 2014. "Preparation of functionalized water-soluble photoluminescent carbon quantum dots from petroleum coke." *Carbon* 78: 480-489. https://doi.org/10.1016/j.carbon.2014.07.029.

Xia, C. L., S. J. Zhu, T. L. Feng, M. X. Yang, and B. Yang. 2019. "Evolution and Synthesis of Carbon Dots: From Carbon Dots to Carbonized Polymer Dots." *Advanced Science* 6 (23). https://doi.org/10.1002/advs.201901316.

Xian, Y., and K. Li. 2022. "Hydrothermal Synthesis of High-Yield Red Fluorescent Carbon Dots with Ultra-Narrow Emission by Controlled O/N Elements." *Adv Mater* 34 (21): e2201031. https://doi.org/10.1002/adma.202201031.

Xie, Ning, Long Tan, Hua-Fei Li, Hai-Yan Hu, Chen Wang, Ming Pan, Fan Wu, Ping Wu, Xiao-Di Wang, Zheling Zeng, Shuguang Deng, and Gui-Ping Dai. 2020. "Manipulation of 3D nanocarbon hybrids toward synthesis of N-doped graphene quantum dots with high photoluminescence quantum yield." *Journal of Luminescence* 219. https://doi.org/10.1016/j.jlumin.2019.116827.

Xu, X. Y., R. Ray, Y. L. Gu, H. J. Ploehn, L. Gearheart, K. Raker, and W. A. Scrivens. 2004. "Electrophoretic analysis and purification of fluorescent single-walled carbon nanotube fragments." *Journal of the American Chemical Society* 126 (40): 12736-12737. https://doi.org/10.1021/ja040082h.

Yamato, K., R. Sekiya, M. Abe, and T. Haino. 2019. "Separation of Spectroscopically Uniform Nanographenes." *Chemistry-an Asian Journal* 14 (10): 1786-1791. https://doi.org/10.1002/asia.201801632.

Yan, X., X. Cui, and L. S. Li. 2010. "Synthesis of Large, Stable Colloidal Graphene Quantum Dots with Tunable Size." *Journal of the American Chemical Society* 132 (17): 5944-+. https://doi.org/10.1021/ja1009376.

Yang, G. C., C. L. Wu, X. J. Luo, X. Y. Liu, Y. Gao, P. Wu, C. X. Cai, and S. S. Saavedra. 2018. "Exploring the Emissive States of Heteroatom-Doped Graphene Quantum Dots." *Journal of Physical Chemistry C* 122 (11): 6483-6492. https://doi.org/10.1021/acs.jpcc.8b01385.

Yang, S. W., X. L. Wang, P. He, A. L. Xu, G. Wang, J. L. Duan, Y. Q. Shi, and G. Q. Ding. 2021. "Graphene Quantum Dots with Pyrrole N and Pyridine N: Superior Reactive Oxygen Species Generation Efficiency for Metal-Free Sonodynamic Tumor Therapy." *Small* 17 (10). https://doi.org/10.1002/smll.202004867.

Yang, Yongsheng, Changxing Wang, Siyuan Tang, Da Chen, Xiameng Li, Zhao Peng, and Chenjie Gu. 2021. "Multicolor Graphene Quantum Dots via Solvatochromic Tuning and Sulfur Doping for Light-Emitting Diodes." *ACS Applied Nano Materials* 4 (11): 12325-12334. https://doi.org/10.1021/acsanm.1c02812.

Ye, R. Q., C. S. Xiang, J. Lin, Z. W. Peng, K. W. Huang, Z. Yan, N. P. Cook, E. L. G. Samuel, C. C. Hwang, G. D. Ruan, G. Ceriotti, A. R. O. Raji, A. A. Marti, and J. M. Tour. 2013. "Coal as an abundant source of graphene quantum dots." *Nature Communications* 4. https://doi.org/10.1038/ncomms3943.

Yin, X. J., C. W. Zhi, W. W. Sun, L. P. Lv, and Y. Wang. 2019. "Multilayer NiO@Co3O4@graphene quantum dots hollow spheres for high-performance lithium-ion batteries and supercapacitors." *Journal of Materials Chemistry A* 7 (13): 7800-7814. https://doi.org/10.1039/c8ta11982a.

Zeng, Q. S., T. L. Feng, S. Y. Tao, S. J. Zhu, and B. Yang. 2021. "Precursor-dependent structural diversity in luminescent carbonized polymer dots (CPDs): the nomenclature." *Light-Science & Applications* 10 (1). https://doi.org/10.1038/s41377-021-00579-6.

Zhang, R. Z., J. R. Adsetts, Y. T. Nie, X. H. Sun, and Z. F. Ding. 2018. "Electrochemiluminescence of nitrogen- and sulfur-doped graphene quantum dots." *Carbon* 129: 45-53. https://doi.org/10.1016/j.carbon.2017.11.091.

Zhang, Wenbo, Li Li, Sichun Li, Jianzhong Ma, Chao Liu, and Yan Bao. 2022. "Modification and application of graphene quantum dots." *Acta Materiae Compositae Sinica* 39 (7): 3104-3120.

Zhang, Z. P., J. Zhang, N. Chen, and L. T. Qu. 2012. "Graphene quantum dots: an emerging material for energy-related applications and beyond." *Energy & Environmental Science* 5 (10): 8869-8890. https://doi.org/10.1039/c2ee22982j.

Zhao, Q. L., Z. L. Zhang, B. H. Huang, J. Peng, M. Zhang, and D. W. Pang. 2008. "Facile preparation of low cytotoxicity fluorescent carbon nanocrystals by electrooxidation of graphite." *Chemical Communications* (41): 5116-5118. https://doi.org/10.1039/b812420e.

Zheng, D. X., P. Han, X. K. Wang, W. F. Sun, J. S. Ye, and Y. Zhang. 2021. "Vibrational properties of graphene quantum dots: Effects of confinement, geometrical structure, and edge orientation." *Physical Review B* 104 (8). https://doi.org/10.1103/PhysRevB.104.085430.

Zheng, X. T., A. Ananthanarayanan, K. Q. Luo, and P. Chen. 2015. "Glowing Graphene Quantum Dots and Carbon Dots: Properties, Syntheses, and Biological Applications." *Small* 11 (14): 1620-1636. https://doi.org/10.1002/smll.201402648.

Zhu, S. J., J. R. Shao, Y. B. Song, X. H. Zhao, J. L. Du, L. Wang, H. Y. Wang, K. Zhang, J. H. Zhang, and B. Yang. 2015. "Investigating the surface state of graphene quantum dots." *Nanoscale* 7 (17): 7927-7933. https://doi.org/10.1039/c5nr01178g.

Zhu, S. J., Y. B. Song, J. Wang, H. Wan, Y. Zhang, Y. Ning, and B. Yang. 2017. "Photoluminescence mechanism in graphene quantum dots: Quantum confinement effect and surface/edge state." *Nano Today* 13: 10-14. https://doi.org/10.1016/j.nantod.2016.12.006.

Zhu, S. J., Y. B. Song, X. H. Zhao, J. R. Shao, J. H. Zhang, and B. Yang. 2015. "The photoluminescence mechanism in carbon dots (graphene quantum dots, carbon nanodots, and polymer dots): current state and future perspective." *Nano Research* 8 (2): 355-381. https://doi.org/10.1007/s12274-014-0644-3.

Zhu, S. J., L. Wang, B. Li, Y. B. Song, X. H. Zhao, G. Y. Zhang, S. T. Zhang, S. Lu, J. H. Zhang, H. Y. Wang, H. B. Sun, and B. Yang. 2014. "Investigation of photoluminescence mechanism of graphene quantum dots and evaluation of their assembly into polymer dots." *Carbon* 77: 462-472. https://doi.org/10.1016/j.carbon.2014.05.051.

Zhu, S. J., J. H. Zhang, C. Y. Qiao, S. J. Tang, Y. F. Li, W. J. Yuan, B. Li, L. Tian, F. Liu, R. Hu, H. N. Gao, H. T. Wei, H. Zhang, H. C. Sun, and B. Yang. 2011. "Strongly green-photoluminescent graphene quantum dots for bioimaging applications." *Chemical Communications* 47 (24): 6858-6860. https://doi.org/10.1039/c1cc11122a.

Chapter 2

Graphene Quantum Dots for Detection - From Inorganic Ions via Biomolecules to Pathogenic Microorganisms

Svetlana Jovanović *
and Sladjana Dorontić
Vinča Institute of Nuclear Sciences, University of Belgrade,
National Institute of the Republic of Serbia, Belgrade, Serbia

Abstract

Graphene quantum dots (GQDs) show exceptional optical, photoluminescent, chemiluminescent, and electroluminescent properties that made them excellent materials for sensing. In this chapter, we summarized the knowledge about GQDs and explored their applications in the detection of inorganic cations and anions, organic molecules, such as pesticides, cancer biomarkers, metabolites like glucose, L-glutamine, and in the detection of enzymes, microorganisms, etc. The methods for detection as well as mechanisms involved in the monitoring of analytes are explained in depth.

Keywords: graphene quantum dots, photoluminescence, sensors, optical detection, pollutants, metabolites

* Corresponding Author's Email: svetlanajovanovicvucetic@vin.bg.ac.rs.

In: Graphene Quantum Dots (GQDs)
Editor: Aleksey Kuznetsov
ISBN: 979-8-89113-638-0

1. Introduction

1.1. Graphene Quantum Dots – Let Us Start with the Structure

Graphene, graphene oxide, carbon nanotubes, carbon quantum dots, graphene quantum dots, nanodiamond, fullerenes, nano onions, carbon nanofibers, nanobuds, carbon nanohorns are all members of a new group of materials named carbon nanomaterials (Cha, Shin et al. 2013, Arnault 2021, Ghalkhani, Khosrowshahi et al. 2021, Prekodravac, Kepić et al. 2021). One feature that they all possess is carbon as the main building element. They are the subject of interest for scientists from many different fields: energy, ecology, medicine, pharmaceutical science, (Lin and Zhang 2012, Cha, Shin et al. 2013, Shan, Zhao et al. 2017, Iqbal, Khatoon et al. 2019) etc.

Graphene quantum dots (GQDs) are one of the youngest members of this group: they were discovered in 2008 (Ponomarenko, Schedin et al. 2008), while fullerene, the oldest member, was found in 1985 (Kroto, Heath et al. 1985).

Due to their structure, GQDs are dispersible in water (Lin and Zhang 2012) and polar organic solvents (Wolk, Rosenthal et al. 2018). In the core of GQDs are sp^2 hybridized carbon atoms organized in 6-membered rings named benzene rings. This part of the GQD structure is the same as the graphene structure and it is hydrophobic. But, GQDs are water-soluble. How is that possible then? Well, numerous oxygen-containing functional groups such as carboxyls, carbonyls, epoxy groups, ethoxy groups, and hydroxyls are found in GQDs. These groups make them water-soluble and subsequently biocompatible. Functional groups are not equally distributed over the GQDs surface: some of them are dominantly located at GQDs edges such as carboxyls and carbonyls, while OH and epoxy groups are usually placed on the GQD core, as presented in Figure 1. When graphene quantum dots contain only C, O, and H elements in their structure, they are considered as non-doped.

Due to a large number of different functional groups and solubility, GQDs have rich chemistry, that can be easily modified which leads to changes in their optical and other properties (Tian, Tang et al. 2018). There are many different approaches to changing their structure such as chemical reduction (Feng, Zhao et al. 2014), amidation (Yanyan, Lin et al. 2021), gamma irradiation (Jovanovic, Syrgiannis et al. 2015, Jovanović, Marković et al. 2016, Jovanović 2021, Milenković, Mišović et al. 2021), thermal annealing (Jovanović, Marković et al. 2017), and many others (Luo, Ji et al. 2013, Chen, Zhao et al. 2018, Yu, Ma et al. 2020, Keshipour, Nejati et al. 2021).

The structure of GQDs is also the main reason for their extraordinary optical properties such as photoluminescence (PL) in the visible part of the spectrum, resistivity to photobleaching, tunable and controllable PL (Thangadurai, Manjubaashini et al. 2022). These features are mainly the results of GQD small size, tunable surface edges, quantum confinement, and edge effects.

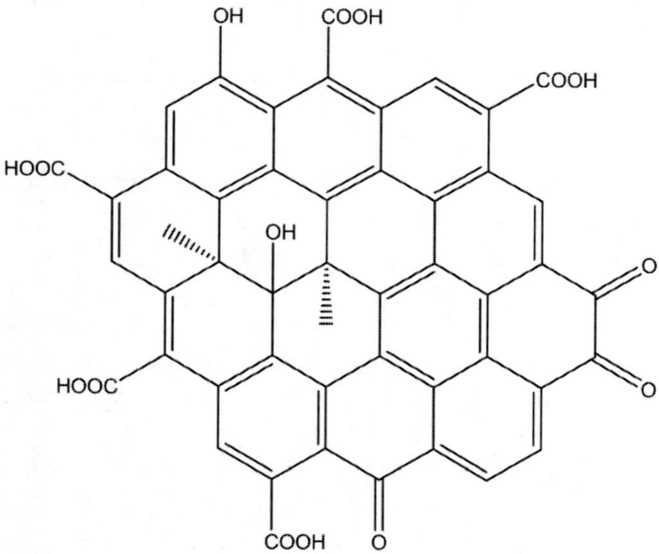

Figure 1. Structure of GQDs. The figure was adapted from Materials 2020, 13(1), 96 (Mansuriya and Altintas 2020). Copyright MDPI 2022.

1.2. Properties of Graphene Quantum Dots

Broad scientific attention is focused on GQDs due to their unusual physical and chemical properties such as water solubility, high surface area, superior chemical inertness, photoluminescence, surface grafting, long-term resistance to photobleaching, and biocompatibility (Li, Liu et al. 2017, Tabish, Scotton et al. 2018).

There exist a large number of studies where possible applications of GQDs were analyzed is related to their photoluminescence (Zou, Gu et al. 2015, (Lu, Li et al. 2019), Shen, Chen et al. 2021(Henna and Pramod 2020, Younis, He et al. 2020, Biswas, Islam et al. 2021, Chung, Revia et al. 2021, He, Hu et al. 2022)). They have been studied as agents for bioimaging and as sensors for many different analytes (Zor, Morales-Narváez et al. 2015, Schroeder, Goreham et al. 2016, Zhu, Liu et al. 2016, Abbas, Tabish et al. 2020, Jovanović, Dorontić et al. 2020, S. A and P. B. C 2020, Dorontić, Jovanović et al. 2021, Milenković, Mišović et al. 2021).

Photoluminescence of GQDs is usually centered in the visible part of the spectrum, but it can extend up to the near-infrared (Zhu, Song et al. 2017). The mechanism of GQD PL was explained by:

- the size of the GQDs, where graphene fragments show quantum confinement effects, a non-zero bandgap (Geim and Novoselov 2007), and PL;
- surface functionalization, which changes bandgap (Jin, Kim et al. 2013);
- doping atoms (Zhu, Song et al. 2017).

Even today an agreement on the PL mechanism has not been achieved. It is most probably the result of the quantum confinement effect of conjugated π-domains, the surface/edge states in GQDs, and the combination of these two mechanisms (Zhu, Song et al. 2017). This property of GQDs was explored for their application in sensing different ions and molecules. The values of PL quantum yield for GQDs have been measured to vary from 2% (Jovanovic, Syrgiannis et al. 2015) to 52.4% (Chen, Lu et al. 2020), or even 81% (Huang, Zhou et al. 2020).

Another interesting feature of GQDs is their ability to produce reactive oxygen species upon illumination in the presence of oxygen molecules (Markovic, Harhaji-Trajkovic et al. 2011, Ristic, Milenkovic et al. 2014, Jovanovic, Syrgiannis et al. 2015, Du, Wang et al. 2016, Kuo, Chang et al. 2016, Jovanović, Marković et al. 2017, Marković, Jovanović et al. 2018, Zhang, Yang et al. 2018, Jovanović, Syrgiannis et al. 2020). This behavoir of GQDs was explored for therapy of carcinoma and bacterial infestions (Markovic, Jovanovic et al. 2018, Monroe, Belekov et al. 2019, Chen, Wu et al. 2020, Santos, Rodríguez-Pérez et al. 2021).

GQDs are also studied for their application in the energy storage and conversion, catalysis, in biomedicine for drug delivery (Dorontić, Jovanović et al. 2021(Iannazzo, Pistone et al. 2017, Biswas, Islam et al. 2021, Liu, Zhao et al. 2021).

In this chapter, the focus will be only on the GQDs application in the detection of metallic ions, organic and biological molecules.

2. Sensing Applications of GQDs

Sensors are devices that are able to detect and/or measure the concentration of molecules or ions and to convert it in a response – signal. It was found that GQDs are good sensing material thanks to high electron motion and high-speed reaction (Jin, Feng et al. 2020, Bressi, Ferlazzo et al. 2021). These features make GQDs a good candidate for sensing devices. A few different types of sensors based on GQDs have been examined:

- optical sensors (Anas, Fen et al. 2019);
- electrochemical sensors (EC) (Facure, Schneider et al. 2021);
- eectrochemiluminescence (ECL) sensors (Kumar, Deshmukh et al. 2020).

Optical sensors are working on different transduction mechanisms, absorption, resonance, or photoluminescence (De Acha, Elosua et al. 2017). The PL sensors are photoluminescent materials made of molecules that change the intensity of their photoluminescent emission signal in the presents of analytes (De Acha, Elosua et al. 2017). GQDs are mostly studied as PL sensors that are able to modulate the intensity of the emission spectra in the presence of the analytes in two ways (Li, Wu et al. 2014, Niu, Zhong et al. 2018, Nair, Thomas et al. 2020, S. A and P. B. C 2020, Sharma, Ali et al. 2021):

1. the increase of intensity or so-called "turn-on" mechanism or
2. the lowering of the signal which is named "turn-off."

It is possible to establish linear relation between analyte concentration and PL emission spectra intensity. The quenching efficiency of the sensor can be evaluated using the Stern-Volmer equation: $A_0/A= 1+ K_{sv}\ [Q]$, where A_0 and A correspond to integrated areas under the PL intensity curve in the absence and the presence of the analyte, respectively (Tam and Choi 2018, Shahat, Elamin et al. 2022). The [Q] is a concentration of the quencher, while K_{sv} is a Stern-Volmer constant that quantifies the quenching efficiency of the sensor.

GQDs were also studied as electrochemical sensors due to their ability to act as a multivalent redox species using cyclic voltammetry (CV) and differential pulse voltammetry (DPV) (Shinde and Pillai 2013). The basic components of electrochemical sensors are:

1. *a molecular recognition system* which is the essential part of a sensor,

2. *a physicochemical transducer system* that transforms the chemical or biological response into a measurable signal that is detectable by electrical devices.

These two components are used to build a working electrode (Faridbod, Gupta et al. 2011). Transduction of a reaction on the working electrode into an electrical signal can be achieved either by voltammetry, impedimetry, amperometry, conductometry, or potentiometry.

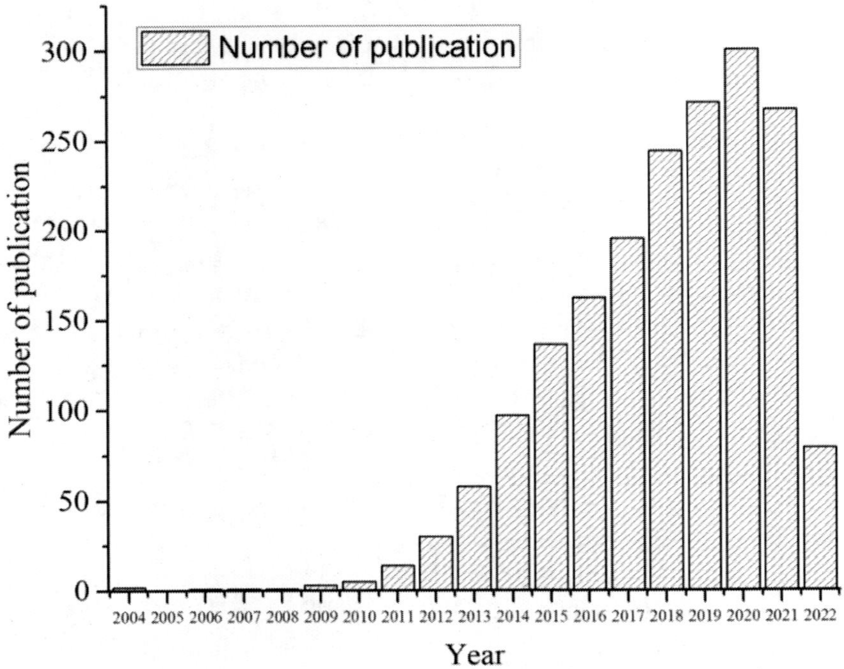

Figure 2. Overview of the number of publications on the GQDs application in sensing. Data were collected in March 2022 from the Scopus database.

Electrochemiluminescence sensors combine both luminescence and electrochemistry analysis (Zanut, Fiorani et al. 2020). After the potential is applied onto an electrode, its surface is excited, and light is emitted as a result of an electron transfer between surface molecules. Lately, a large scientific interest has been focused on the application of GQDs in electrochemiluminescence sensing, particularly in biosensing (Liang, Qiu et al. 2016, Zhou, Chen et al. 2017, Vermisoglou, Panáček et al. 2020).

When the Scopus database was used to find the research articles using keywords "graphene quantum dots" and "detection," the results were 1866 documents. Figure 2 presents the dynamic of the number of articles published in this area. It is undoubtedly that GQDs have been attracting huge scientific attention in the last 5 years in the field of detection.

In the following part of the chapter, the possibilities of GQDs application in the detection of different analytes are discussed.

2.1. Detection of Inorganic Ions

GQDs have been studied as a sensor for many different cations and anions using various approaches. Haung et al. have first noticed the ability of transition metals ions to quench photoluminescence of GQDs and they associated this behavior with metal-ion electron configuration (Huang, Liao et al. 2013). They observed that if d-orbitals of metal ions are empty or completely filled, the ions are unable to quench the GQD photoluminescence, while the ions with partially filled d-orbitals efficiently lowered PL of GQDs. In the case of ions with empty or completely filled d-orbitals, a closed-shell electron configuration is achieved; therefore, they are in a stable energy state. This is a reason why these kinds of ions do not establish electron transfer with GQDs and therefore have no significant effect on the GQD PL.

Metal ions interact with GQDs in 2 modes:

- through O-containing functional groups by chelating (Kappen, Bharathi et al. 2022);
- through the π-system (electron cloud) of graphene structure in GQDs (Park, Lee et al. 2008).

Metal ions with semi-filled d-orbitals accept electrons from the π-system of GQDs, due to the ability of GQDs to serve as the electron donor in the reaction with electron-withdrawing reagents (Barone, Baik et al. 2005). When GQDs establish electron-donor interactions with metal ions, it causes the perturbations of electronic states in dots and thus non-radiative transitions, leading to the reduction of the GQDs photoluminescence.

Nitrogen-doped GQDs were produced by hydrothermal treatment of GQDs with hydrazine and used for detection of Fe^{3+} ions (Ju and Chen 2014). Detection was accomplished using GQDs as a fluorescence probe and they showed sensitivity in a wide concentration range of 1-1945 μM with a limit of detection (LOD) of 90 nM.

GQDs were also studied as a fluorescence probe for the detection of Cu^{2+} in water (Wang, Gu et al. 2014). It was established that the intensity of emission spectra was inversely proportional to the concentration of Cu^{2+} ions. The linear ion concentration range was between 0 and 15 μM, with the LOD of 0.226 μM. Pristine GQDs (Liu, Gao et al. 2014) were used for the detection of Cu^{2+} ions. Dots were produced using graphite fibers that were chemically oxidized. It was shown that photoluminescence (PL) of the as-synthesized dots was quenched by Cu^{2+} ions and recovered with biothiolcysteine. A linear range was from 0 to 0.20 mM with the LOD of 0.33 μM. Amino-functionalized GQDs were also used in Cu^{2+} ions detection based on the fluorescence quenching with the LOD of 5.6 nM (Tam and Choi 2018).

Detection of heavy metal ions such as Pb^{2+} was also achieved using GQDs (Zhou, Jiang et al. 2014). The electrochemical method was used for the determination of Pb^{2+} ions with the LOD as low as 7×10^{-9} M. This ion was also detected using P-, N-, S-doped GQDs (Xu, Wang et al. 2018). A linear concentration range was between 1 and 20 μM with the LOD of 0.75 μM. The mechanism of detection was established to be the fluorescence quenching. Pb^{2+} ions were also detected with GQDs functionalized with dithienopyrrole derivative (DTPAN) (Sebastian, Kala et al. 2021). Detection was based on the fluorescence quenching with a linear concentration range from 0.5 to 40 nM and the ultralow LOD of 0.25 nM.

N-doped GQDs were used in the ferric ions detection (Li, Li et al. 2015). The PL of the doped GQDs was quenched by both Fe^{3+} and Hg^{2+}. A linear concentration range was found to be between 0.5 μM and 0.5 mM of Fe^{3+} with the LOD of 0.1 μM.

When GQDs were conjugated with gold nanoparticles (AuNPs), a new system for the electrochemical detection of Hg^{2+} and Cu^{2+} was produced (Ting, Ee et al. 2015). This study showed the ultralow LODs, 0.02 nM for Hg^{2+} and 0.05 nM for Cu^{2+}. N-doped GQDs were also found to be able to detect Cu^{2+} ions with the LOD of 57 nM (Gao, Chen et al. 2020). Gamma irradiated amino-functionalized GQDs showed the LOD of 0.189 nM/L and the linear concentration range from 0.3-16.1 μM/L for the detection of Cu^{2+} ions (Jovanović, Dorontić et al. 2020).

Europium-decorated GQDs were studied as "off-on" photoluminescent sensor for the detection of Cu^{2+} and L-cysteine (Lin, Song et al. 2015). The linear range of 0.1-10 μM and LOD of 0.056 μM for Cu^{2+} and the linear range of 0.5-50 μM and LOD of 0.31 μM for L-Cys were reported.

Multiple detections of Cu^{2+}, Pb^{2+}, and As^{3+} were achieved by GQDs produced by cage-opening oxidation of C_{60} buckminsterfullerene (Ciotta, Paoloni et al. 2017). The mechanism of the detection was turn-off fluorescence sensing, while the changes in emission were assigned to the aggregation which may arise from multiple complexations, where ions connect with few GQDs, or from chelating complex formation with carboxyl or hydroxyl groups at GQDs, or the aggregation might be a result of edge-to-edge geometry as suggested by Figure 3.

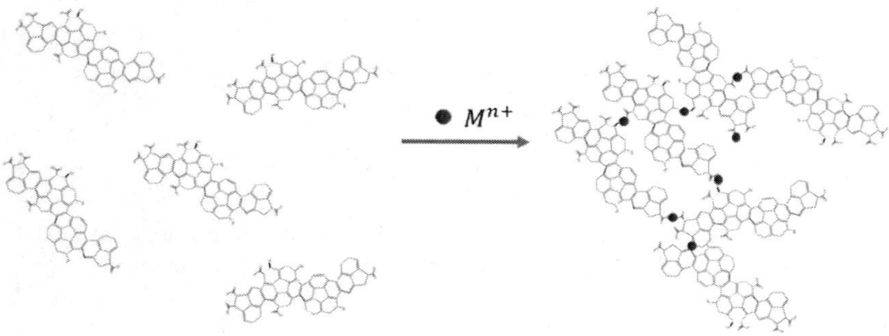

Figure 3. Schematic presentation of complex formation between metal ions and GQDs. Figure adapted from Sensors 2017, 17(11), 2614 (Ciotta, Paoloni et al. 2017). Copyright 2022, MDPI AG.

Investigated GQDs showed high selectivity for the detection of Cu^{2+}, Pb^{2+}, and As^{3+} (Figure 4).

Nonetheless, the explanation of the distinct behavior of transmittance in the presence of As^{3+} and the formation of precipitate with Pb^{2+} require further studies, due to lack of interaction mechanism. These differences, however, also make it possible to discriminate between the three metal ions in the view of the implementation of selective multiple sensors.

Furthermore, Ag^{+} ions were detected using N-doped GQDs (Tabaraki and Nateghi 2016), where the linear range was found to be between 0.2 and 40 μM with the LOD of 168 nM. Detection of Ag^{+} in biological samples was achieved using GQDs functionalized with DNA by green fluorescence quenching (Mandal, Khatun et al. 2019).

Au^{3+} ions were also detected using GQDs (Amjadi, Shokri et al. 2016). Turn-off fluorescence sensor based on GQDs was showed to have a linear response in the concentration range of 1.0-80 μM, with the LOD of 0.5 μM.

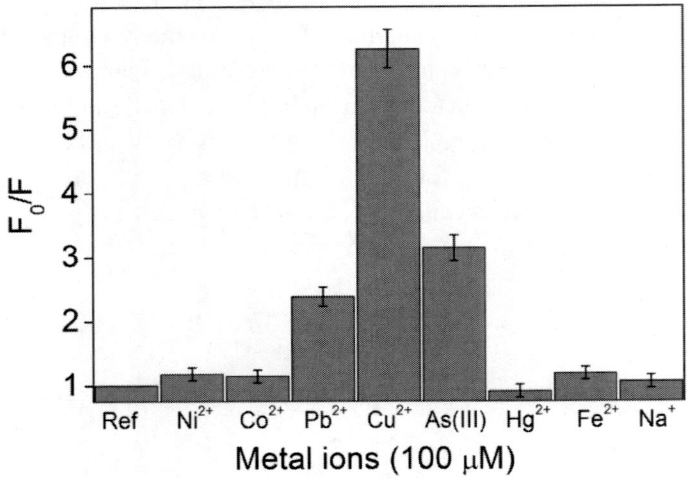

Figure 4. The ratio F_0/F (where F_0 and F correspond to integrated areas under the PL intensity curve in the absence and the presence of the metal ions) in the presence of different metal ions at the same concentration. Figure adapted from Sensors 2017, 17(11), 2614 (Ciotta, Paoloni et al. 2017). Copyright 2022, MDPI AG.

Mercury ions Hg^{2+} were detected using S,N-co-doped GQDs (Anh, Chowdhury et al. 2017). Synthesized quantum dots were 3.5 ± 0.5 nm in the lateral size and had a high PL quantum yield, 41.9%. Incorporated sulfur atoms improved the coordinative bonding of Hg^{2+}. The mechanism of detection was the turn-off fluorescence sensing, with the linear range from 0.1 to 15 μM in both sewage and dye wastewater. N-doped and S,N-codoped GQDs were also used to study the Hg^{2+} detection (Mao, Zhao et al. 2019, Qu, Zhang et al. 2019). In both cases, the turn-off fluorescence mechanism was found to be behind the sensing. Ultralow concentrations (the LOD of 10 ppb) of Hg^{2+} ions were detected with the N-doped GQDs (Fu, Hsieh et al. 2020).

GQDs with a record PL quantum yield of 99.8% were produced by the pyrolysis of citric acid with D-penicillamine (DPA) or L-penicillamine (LPA) (Xuan, Ruiyi et al. 2017). A turn-off fluorescence sensor for the detection of Fe^{3+} ions was produced with the linear concentration range from 0.004 to 1.8 mM and the LOD of 1.2×10^{-6} M. The mechanism of interaction between dots and ions was found to be based on coordinative bonding. The same ion was detected using boron-doped GQDs (Chen, Wu et al. 2017). The linear concentration range was found to be very wide, from 0.01 to 100 μm with the LOD of 0.005±0.001 μM. It was found that the GQDs were able to recognize Fe^{3+} ions in the acidic solution (Zhu, Zhang et al. 2017). Fe^{3+} ions create coordinative bonds with the OH groups from the GQD structure and induce aggregate formation and consequently fluorescence quenching.

Blue photoluminescent S,N-co-doped GQDs with a very high PL quantum yield (QY), 67%, were produced by pyrolysis of citric acid, as a C source, and amino acid cysteine as a source of heteroatoms (Chen, Zhao et al. 2017). These dots were shown to act as a dual fluorescent and AgNP-related colorimetric system for the detection of cyanide ions (CN^-). The LOD for fluorescent detection was found to be 0.52 μM and 0.78 μM for colorimetric measurement.

Heavy metal ions such as cadmium (II) and lead (II) were detected using integrated glycidyl methacrylate functionalized fluorescent graphene quantum dots and N,N'-methylenebisacrylamide co-cross-linked poly(thioctic acid) with physical cross-linked fluorescent gelatin (Lai, Jin et al. 2022). Herein, GQDs were covalently fixed to into the hydrogel. This gel showed the ability to change the color after the absorption of Cd^{2+}/Pb^{2+} ions from cyan to chartreuse, while fluorescence changed as well and showed a turn-on/turn-off shift. This gel showed the capacity to bind both ions from water with an efficiency of 194.553 mg/g for Cd^{2+} and 259.067 mg/g for Pd^{2+} ions. Other GQDs for fluorescence turn-on sensing of Pd^{2+} ions were produced (Niu, Zhong et al. 2018). A DNA chain named catalytic strand used a linker to link GQDs and AuNPs. When GQDs were bound to AuNPs, fluorescence was quenched and recovered by adding Pd^{2+} ions. A wide range of linear responses was achieved, from 50 nM to 4 μM, with the detection limit of 16.7 nM.

Detection of H_2O_2 without enzymes was achieved using GQDs as well (Zhu, Lu et al. 2017). An electrochemical sensor based on GQDs, polydopamine (PDA) that served as an adhesive agent, and gold nanoparticles (Au NPs) showed an exceptional electrocatalytic activity for hydrogen peroxide reduction. A linear concentration range was found to be wide, 0.1–40 μM with the LOD of 5.8 nM.

Both Hg^{2+} ions and biothiols were detected using the GQDs based sensor (Achadu and Nyokong 2017). For the sensor, a complex nanomaterial was produced: polyethyleneimine-functionalized GQDs (PEI-GQDs) with mercaptopyridine-substituted zinc phthalocyanine (Pc)-Au@Ag nano-alloys. Both analytes were studied in water solution, where Hg^{2+} ion caused quenching and thiols induced the restoring of photoluminescence due to strong metal-thiol bond formation. Hg^{2+} ions were detected using N-doped GQDs with the complexation reaction between porphyrin and manganous ions, Mn(II) (Peng, Zhang et al. 2018). The mechanism of detection was based on a substitution reaction where the porphyrin nucleus was deformed by larger Hg(II) ions. In this situation, the reaction with Mn(II) carried by NGQDs is favorable. Photoluminescence of the porphyrin was quenched when metalloporphyrin is formed, while PL intensity of N-GQDs is increased due to the inner filter effect between porphyrins and N-GQDs. Similarly, N-doped GQDs were found to have the LOD between 0.45 and 67.3 nM (Yang, Xiao et al. 2019). Rhodamine B was found to assist in Hg^{2+} ions detection when N-doped GQDs were produced using both Rhodamine B and ethylenediamine as a nitrogen source (Yang, Xiao et al. 2019), while S,N-doped GQDs also showed good potential in detection of this ion (Gu, Hsieh et al. 2019).

By employing Density Functional Theory (DFT) calculations, binging energies between the GQDs model and Cd^{2+}, Hg^{2+} and Pb^{2+} were found to be 11.66, 14.56, and 7.05 eV (Shtepliuk, Caffrey et al. 2017).

One more modification of GQDs showed to be an excellent approach to improve their ability to detect Fe^{3+} ions (Gao, Zhou et al. 2018). N,Fe-co-doped GQDs were produced by the hydrothermal method. They showed a high sensitivity in the range of 10-110 μM as detected by fluorescence, with the very low LOD (3.21 μM).

Both sulfide ions and ascorbic acid (AA) were detected using the GQDs sensor (Na, Qu et al. 2018). First, GQDs were modified with sulfanilic acid and glutathione. In this way, sulfo- and thiol-terminated groups were introduced at the GQD structure which led to green photoluminescent emission. Due to the presence of sulfogroups, GQDs were more stable in a wide pH range, while thiol groups improved the capability of the dots to bind Cu^{2+}. Also, sulfanilic acid and glutathione modified GQDs-Cu(II) complex was shown to be a redox-active fluorescent sensor for the ascorbic acid detection. One more sensor for the AA detection was based on N-doped GQDs forming the "signal-off" and "signal-on" systems (Novoa-De León, Johny et al. 2019).

Simple and fast detection of Hg^{2+} and Fe^{3+} ions was achieved using photoluminescence and paper test tracks with GQDs (Wang, Sun et al. 2018). Hg^{2+} ions were detected in 3 s, while Fe^{3+} ions were detected in 2 minutes, with the low detection limit of 0.1 μM in water. For the detection, N-doped GQDs were produced starting from citric acid and glycine by hydrothermal treatment at 200°C for 5 h. They were placed at PVA paper and showed the ability to change the color in the presence of selected ions. Fe^{3+} ions were detected using the N-doped GQDs (Lu, Zhou et al. 2019). Herein the LOD was found to be 2.37 μmol/L and the linear range was shown to span from 1600 to 6000 μmol/L. Both Fe^{3+} and Cr^{3+} were detected using GQDs with the LODs of 1 and 3 μM, respectively (Raj, Rajput et al. 2019). N-doped GQDs with Mn^{2+} ions served also as a sensor for the Hg^{2+} detection (Yang, Qin et al.

2018). A linear concentration range was found to vary from 0 to 3.5×10^{-6} mol L^{-1} with the LOD of 3.4×10^{-8} mol L^{-1}. When GQDs were functionalized with folic acid, they were able to selectively bind to tumor cells through folate receptors allowing selective fluorescence detection of Hg^{2+} with the linear range of $2.0 \times$ 10-6-5.0×10^{-12} M and LOD of 1.7×10^{-12} M (Li, Wang et al. 2018).

Both Fe^{3+} and Hg^{2+} were also detected using N,S-GQDs (Yang, Zou et al. 2019). Employing density functional theory (DFT) simulations and comparing their results with experimental data, the inner filter effect (IFE) and dynamic quenching process were proved to act as the key factors in the fluorescence quenching and detection of ions. The incorporation of N and S atoms in the GQD structure leads to charge redistribution, which results in formation of new negative charge sites. Figure 5 presents how metal ion is incorporated in the GQDs structure, inside of a heterocyclic ring. The main force is an electrostatic attraction between positively charged ions and negatively charged rings. The limits of detection for Fe^{3+} and Hg^{2+} were found to be 2.88 and 0.27 nM, respectively.

As^{3+} ions were also detected using GQDs (Pathan, Jalal et al. 2019). The authors showed that magnetic Fe-containing GQDs are efficient fluorescence quenching sensors for this ion with a good selectivity at different pH values. Obtained LOD was shown to be 5.1 ppb.

Uranium (VI) ions detection was achieved with GQDs as well (Guin, Ambolikar et al. 2018). Two mechanisms of interactions between dots and U were proposed:

- Förster resonance energy transfer involving f-orbitals of U (VI) ion;
- nonspecific electrostatic association between quantum dots and U ions.

These interactions induced fluorescence quenching of GQDs and the detection of U at a concentration of 0.56 μg/L. Another study showed a possibility for detection of U(IV) ions, with the LOD of 1.5 ppb (Dewangan, Khan et al. 2019), where the mechanism of detection was the fluorescence quenching as well.

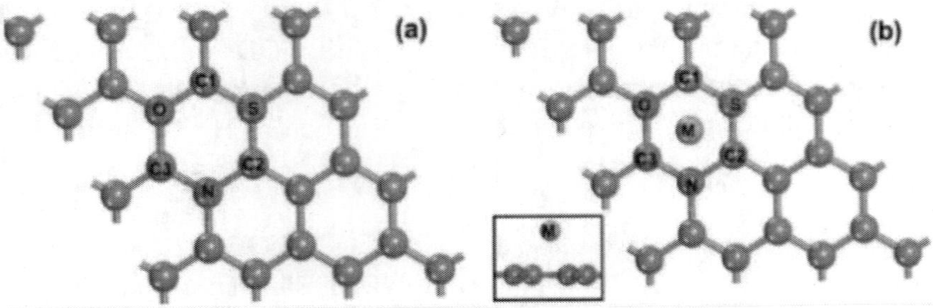

Figure 5. A heterocyclic ring inside GQDs (a) and its interaction with a metal ion (b). Figure adapted from Nanomaterials 2019, 9(5), 738 (Yang, Zou et al. 2019). Copyright 2022, MDPI AG.

Fe^{3+} ions were detected using GQDs produced from watermelon rids (Rodwihok, Tam et al. 2022). Apart from hydroxyl and carboxyl groups, GQDs had amide functional groups as well and showed a prominent blue emission. Fluorescence quenching was the mechanism of the ion sensing and the LOD of 0.28 μM was achieved. In Figure 6a, the results of the selectivity tests show that only Fe^{3+} ions induced a large change in the PL intensity while other ions did not cause photoluminescence quenching. In Figure 6b, the correlation between the concentration of Fe^{3+} ions and ratio F_0/F (F_0 is PL intensity of GQDs, while F is PL intensity of GQDs in the presence of Fe^{3+} ions at indicated concentration) are shown. The same ions were detected with yellow N-doped GQDs, with the linear concentration range of 0–80 μM and the LOD of 63 nM (Wang, Chen et al. 2020).

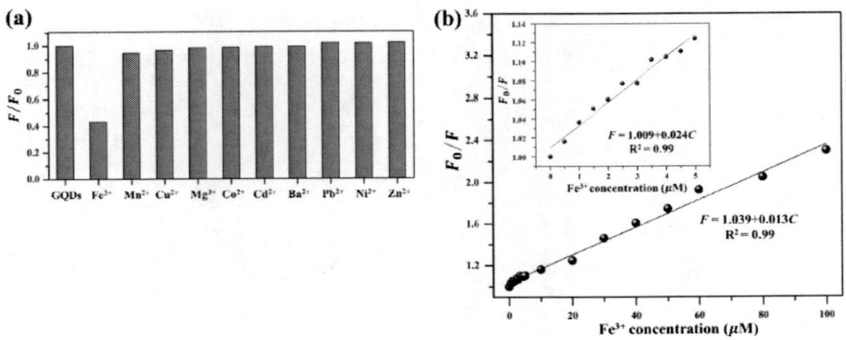

Figure 6. The ratio F_0/F of GQDs in the presence of different ions (a) and the values of ratio F_0/F at different concentrations of Fe^{3+} ions. Figure adapted from Nanomaterials 2022, 12(4), 702 (Rodwihok, Tam et al. 2022). Copyright 2022, MDPI AG.

The low detection limit for Ca^{2+} (2 pM) was achieved using GQDs with crown ether structure that was suitable for both *in vitro* sample as well as for the detection of ions in cells (Huang, Zhou et al. 2020).

2.2. Detection of Organic Molecules

Detection of organic molecules, essentially those that are toxic and hazardous, is an extremely important subject and therefore a huge scientific effort was dedicated to find new, fast, simple, eco-friendly, reliable, and economically acceptable sensors. GQDs seem to be a very good candidate for this and in the following part of the chapter we will review the studies that are focused on the detection of organic molecules.

One of the first study that employed GQDs as a sensor for organic molecules was focused on the detection of malachite green (MG) (Hou, Bei et al. 2013). Herein, a glassy carbon electrode was functionalized with GQDs and AuNPs. The electrochemical activity was studied using cyclic voltammetry. A linear relationship was found between the oxidation peak current and the analyte concentration. The sensor showed good linearity from 4.0×10^{-7} to 1.0×10^{-5} mol/L and the LOD was 1.0×10^{-7} mol/L.

Melamine was also detected using the GQD-based sensor (Li, Wu et al. 2014). The fluorescence sensor was based on the charge transfer quenching of GQD photoluminescence in the presence of Hg^{2+} ions. Interaction between GQDs, melamine, and Hg^{2+} ions was established in the following order: Hg^{2+} ion formed coordinative bonds with N-atoms of melamine and pass more Hg^{2+} ions to GQD surfaces, then π-π stacking resulted in the quenching of the GQD PL. The mechanism of the PL quenching was found to be the charge transfer from GQDs to Hg^{2+} ions through melamine. A linear concentration range was shown to be 0.15-20 μM with the LOD of 0.12 μM.

Vitamin C was also detected with a GQDs sensor (Liu, Chen et al. 2015). The presence of Cu^{2+} was needed in order to establish electron transfer between Cu^{2+} and GQDs, which quenched GQDs photoluminescence while adding ascorbic acid was found to recover the PL of GQDs. A linear concentration response was in the range 0.3-10 μM with the LOD of 94 nM.

A composite based on GQDs and gold NPs was used for electrochemical detection of p-acetamidophenol (Xiaoyan, Ruiyi et al. 2016). Herein, proline functionalized GQDs were mixed with Au^{3+} ions and the complex was created. In the next synthetic step, reduction led to AuNPs formation and the GQDs-AuNPs hybrid was produced. Electrochemical detection of p-acetamidophenol was achieved with low detection limit of 0.02 μM, and wide concentration range (0.08 to 100 μM).

One interesting new application of GQDs in the organic compounds' detection is sensing of CaMV35S promoter in transgenic soybean (Li, Sun et al. 2016). A quantitative analysis of DNA using N-doped GQDs was achieved, and Förster resonance energy transfer (FRET) mechanism was used to detect the cauliflower mosaic virus 35s (CaMV35S) promoter of the transgenic soybean. A linear concentration range was reported to be from 0.1 to 500.0 nM and the LOD was 0.03 nM, which indicates a successful production of an ultrasensitive biosensor.

Benzene molecules were successfully detected using a complex sensor based on GQDs (Zhang, Rong et al. 2018). As a sensing material, a boron-doped GQDs with Ag–$LaFeO_3$ p–p heterojunction (B/APPH) was used. It was found that band gaps of B-GQDs and APPH are similar, which improves the carrier transport ability. The Ag particles catalyzed the reaction of benzene oxidation and production of CO_2 and H_2O in gas-sensing reactions.

Polycyclic aromatic hydrocarbons (PAH) with N, S, or O in the structure were detected using GQDs as well (Li, Wang et al. 2018). These pollutants were studied on the epidermis of mangrove roots. The decrease in the intensity of photoluminescence emission spectra was proportional to the amount of N/O/S-containing PAHs in the epidermis. The linear response was found to be 10.3–980 ng/g, 9.5–1350 ng/g, and 7.8–1200 ng/g for the N-, S-, and O-doped PAHs, respectively.

Cysteine was also detected using reduced GQDs coated AuNPs (Hai, Lin et al. 2018). Composite AuNPs@r-GQDs showed absorption band at 525 nm. But cysteine was found to cause the aggregation of AuNPs@r-GQDs and the color change, and this allowed the amino acid detection with the LOD of 5.6 nM.

Polyphenol quercetin was detected using GQDs based electrode electrochemical determination (Li, Qu et al. 2016). The linear response was found to be in the concentration range of 0.01 to 6.0 μM with the LOD of 2.0 nM. Another approach is based on GQDs/AuNPs electrochemical detection with square-wave voltammetry (Stefanov, Negut et al. 2020). A wide linear concentration range from 1.0×10^{-10} to 1.0×10^{-3} mol/L was achieved, along with the LOD of 3.3×10^{-11} mol/L. This method was used to determine quercetin in human plasma.

Organic pollutant 4-nitrophenol was detected in complex water samples using a sensor based on GQDs (Anh and Doong 2018). Herein in only one stage, Au NPs and sulfur-doped GQDs were produced. When 4-nitrophenol was added, the changes in the UV-Vis spectra at 307 and 530 nm were measured, even at the nanomolar level. It was suggested that GQDs and 4-nitrophenol established π-π interactions. The LOD was 3.5 nM when the medium for detection was deionized water, while the linear range of the extremely polluted wastewater was found to be 0.05-50 μM with a higher value of the LOD (8.4 nM). Another paper reporting 4- nitrophenol detection was focused on N-doped GQDs (Wang, Fu et al. 2021). Herein different detection mechanism was employed for the pollutant detection – fluorescence quenching. 4-nitrophenol was detected at the concentration range 0-20 μM, with the LOD of 0.29 μM.

An electrochemical sensor for the monitoring of aflatoxin M1 (AFM1) was also made using GQDs (Shadjou, Hasanzadeh et al. 2018). First, GQDs were deposited on the electrode surface using cyclic voltammetry, then α-cyclodextrin was polymerized using the same technique as for GQD deposition, and AgNPs were synthesized on the surface of the modified electrode using electrochemical deposition. Modified electrodes provided a large surface area for the detection of AFM1. Using cyclic voltammetry, the linear ratio concentration-signal was established when the concentration of AFM varied from 0.015 mM to 25 mM with the LOD of 2 μM. The other aflatoxin B1 (AFB1) was detected by an electrochemical immunosensor using the GQD-AuNP hybrid (Bhardwaj, Pandey et al. 2019). This sensor had linear response in the concentration range from 0.1 to 3.0 ng/mL and was used for analysis of the real maize samples. Aflatoxin B1 was also detected with the GQDs-AuNPs using linear voltammetry and electrochemical impedance spectroscopy (Gevaerd, Banks et al. 2020). Herein detection limit was found to be 0.45 nmol/L and the linearity was in the range from 1 to 50 nmol/L.

The detection of the small organic molecule formaldehyde in the food products using GQD based sensor was reported recently (Zhang, Qi et al. 2021). In Figure 7, the detection mechanism is presented. It is shown how the photoluminescence of the GQDs in the test paper was quenched by formaldehyde released from the meat selected as a sample food. The mechanism of detection was static quenching, the linear range was found to be 0-1 μg/mL while the LOD was 0.0515 μg/mL in water.

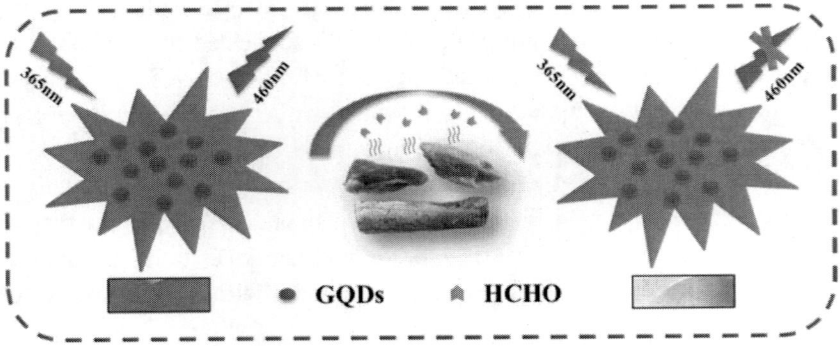

Figure 7. The mechanism of formaldehyde detection in meat with the GQDs test paper. Figure adapted from Sustainability 2021, 13, 5273 (Zhang, Qi et al. 2021). Copyright 2021, MDPI AG.

Hydroquinone detection with electrochemical sensors based on GQDs and copper oxide was reported as well (Chen, Li et al. 2019). A wide linear response range from 0.001 to 40 μM was achieved with the LOD of 0.31 nM.

Cimetidine was detected using chemiluminescence with GQDs and AuNPs as a sensor (Yousefzadeh, Abolhasani et al. 2019). Detection was found to be linear in the concentration range 0.8–200 ng/mL with the LOD of 0.3 ng/mL.

Detection of acetone was also achieved with the GQDs based sensor (Zhang, Wu et al. 2019). A metal-organic frameworks-derived zinc oxide nanopolyhedra/S,N-graphene quantum dots/polyaniline (ZnO/S,N-GQDs/ PANI) ternary nanohybrid was produced on the surface of flexible polyethylene terephthalate (PET) substrate using *in situ* polymerization. The LOD for acetone detection was 0.1 ppm, which is one of lowest reported values.

Bisphenol A (BPA) detection in the living cells is one more astonishing ability of GQDs sensors (Ashraf, Asif et al. 2020). For electrochemical sensing, first single-crystal Cu_2O nanostructures (i.e., Cu_2O-CuO) were produced and covered with GQDs. Obtained material served as an electrode for EC sensing and the LOD of 1 nM was achieved. The sensor showed a wide linear range from 2 nM to 11 mM and was used in the BPA detection in both water and human serum.

Erythrosine B (ErB) is an organic compound from xanthenes groups used in foods, drugs, and cosmetics but it is potentially genotoxic and mutagenic. This is why its concentration in food must be carefully monitored, and the GQDs based sensor showed potential for this application (Farmani, Peyman et al. 2020). Hybrid nanomaterial made of GQDs and AuNPs (GQD-AuNPs) was able to interact with ErB which resulted in the decrease of the PL intensity. The linear concentration range was found to be from 1.2 nM to 50 nM while the LOD was 0.03 found to be nM.

Curcumin was detected using a fluorescence quenching approach with GQDs as a sensor with the remarkable LOD of 30.0 nM/L (Tade and Patil 2020).

Amino acid L-cysteine was detected using gold nanostar and GQDs composite (Abolghasemi-Fakhri and Amjadi 2021). Gold nanostars are multibranched nanoparticles with multiple plasmon resonances, presented in Figure 8 (Chirea 2013). The detection was achieved at the nM level.

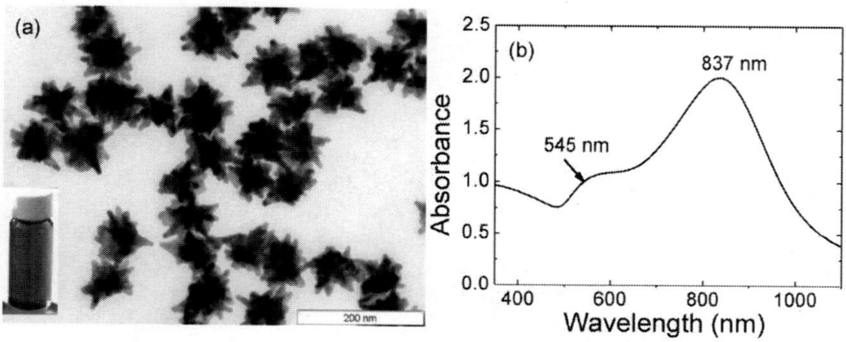

Figure 8. TEM image (a) and optical absorption spectrum of AuNS (b). Figure adapted from the Catalysts 2013, 3(1), 288 (Chirea 2013). Copyright 2022, MDPI AG.

Thiocyanate concentration was measured using a colorimetric approach and AuNPs-GQDs hybrid and 3,3′,5,5′-tetramethylbenzidine (TMB), and H_2O_2 in the system (Rahin Ahmed, Sherazee et al. 2022). Detection was based on OH radicals scavenging activity of SCN^- ions and interactions between sulfur and gold. Due to nanoenzymatic reactions between Au NPs in the presence of TMB and molecules of water peroxide, blue color was changed, allowing colorimetric quantification of SCN^- ions. Ultrasensitive detection of thiocyanate was reported with the LOD of 3 nM.

Apart from organic molecules, GQDs are incorporated in sensors for the detection of a vast number of metabolites, and those sensors are considered in the following section.

2.3. Detection of Metabolites and Antigens with GQDs Sensors

The mechanism of action of the sensors for glucose detection based on GQDs and enzyme reaction are presented in Figure 9 (Wang, Shen et al. 2018).

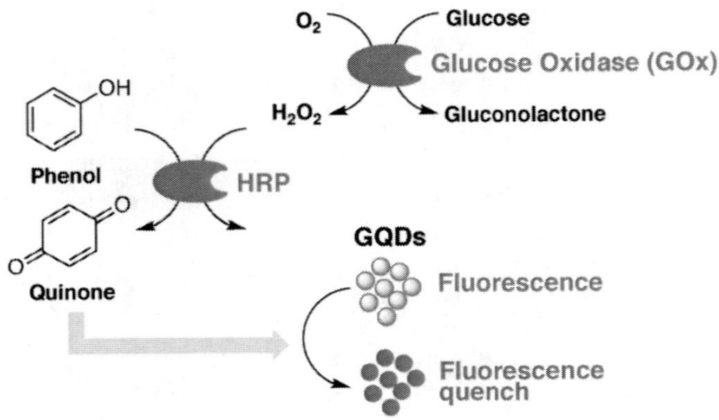

Figure 9. A schematic illustration of a mechanism for the glucose detection with glucose oxidase enzyme (indicated as GOx), horseradish peroxidase enzyme (named HRP), molecules of phenol, and GQDs catalyzed reaction. The figure is adapted from International Journal of Molecular Sciences, 2018, 19(6), 1696 (Wang, Shen et al. 2018). Copyright 2022, MDPI AG.

The molecules of sugar are unable to directly quench the GQDs photoluminescence. Thus, the process of detection was based on the indirect ability of glucose molecules to quench photoluminescence of GQDs in the GOx-HRP-phenol system. First, the enzyme GOx catalyzes the conversion of glucose into gluconolactone. During this reaction, molecules of H_2O_2 are produced as well. Then the enzyme HRP allows the conversion of phenol into quinone by oxidation with H_2O_2. Produced quinone is a good electron acceptor, which is able to remove the electron from the excited GQDs and to cause PL quenching. All components were mixed at once.

The colorimetric approach for the glucose detection was reported by Chen et al. in 2014 (Chen, Hai et al. 2014) using GQDs functionalized with silver nanoparticles (AgNPs). This hybrid material showed a superior absorbance fading response for the hydrogen peroxide reduction and allowed the development of a simple colorimetric procedure for the H_2O_2 detection. The LOD of 33 nM was achieved for H_2O_2, while the glucose detection was accomplished due to the specific catalytic activity of glucose oxidase for the oxidation of glucose and production of H_2O_2. Herein, the LOD was 170 nM proving that ultrasensitive detection of glucose was possible in this way. Another study showed that DNA-mediated AgNPs and GQDs hybrid nanocomposite (AgNP-DNA@GQDs) can be a fluorescent sensor for H_2O_2 (Wang, Zheng et al. 2014). The LOD was reported to be 0.10 μM for the H_2O_2 detection, in the presence of glucose oxidase (GOx) and the glucose in human urine sample (Figure 9). One more sensor for the glucose detection was made using GQDs (Samuei, Fakkar et al. 2017). The complex nanomaterial based on GQDs and CoNiAl-layered double-hydroxide was obtained using the co-precipitation method and used as an electrode paste for the non-enzymatic glucose detection. Cyclic voltammetry and chronoamperometry methods were used to study the electrochemical properties of the material in the presence of glucose. A linear concentration range was reported from 0.01 to 14.0 mM with the LOD of 6 μM.

Ultrasensitive detection of the prostate-specific antigen (PSA) was reported using graphene oxide quantum dots@silver (GOQDs@Ag) core-shell nanocrystals (Ting, Ee et al. 2015). The immunosensor was constructed using magnetic beads (MBs) with immobilized anti-PSA antibody and GOQDs@Ag were a detection probe. The linear relationship between the PL intensity and the PSA concentration was reported in the range from 1 pg/mL to 20 ng/mL with the LOD of 0.3 pg/mL. One more GQDs biosensor for the PSA detection was developed with the linear concentration range of 2–9 pg/mL and LOD of 1.8 pg/mL (Malekzad, Hasanzadeh et al. 2017). In this sensor, PSA-antibodies were placed on the L-cysteine (L-Cys) based matrix and GQDs/AuNPs composite that worked as a signal amplifier. The PSA antigen was detected also using the $Ni(OH)_2$/NGQDs sensor (Zhu, Saddam Khan et al. 2018). Herein, the sensing mechanism was reported to be the quenching of electrochemiluminescence (ECL) based on the ECL resonance energy transfer. Due to excellent ECL emission, N-doped GQDs were used in this sensor. The dots were placed onto $Ni(OH)_2$, in a 3D lamellar structure. $Ni(OH)_2$ was selected due to its good conductivity and large surface area. The second part of the sensor were Fe_3O_4@MnO_2 composites. This material was able to absorb wider range of the ultraviolet light. The sandwich architecture of the immunosensor was applied in the study, and the PSA detection was linear within the concentration range of 10^{-5}–10 ng/mL and with the LOD of 5 fg/mL.

Dopamine was also detected using the GQDs based sensor (Yan, Liu et al. 2015). GQDs-TiO_2 nanocomposite was shown to demonstrate an enhanced photoelectrochemical (PEC) signal when the material was exposed to visible light. The nanocomposite photocurrent was detected in the presence of dopamine and increased linearly with its concentration in the range 0.02 – 105 μM. One more system for the dopamine detection was based on the composite of GQDs and AuNPs (Das, Paul et al. 2019). The dopamine concentration was measured accurately in the range 0–10 μM with the LOD of 40 nM. Titania-ceria-GQDs were used for the dopamine detection as new electrochemical (EC) and photoelectrochemical (PEC) sensors (Ahmadi, Bagherzadeh et al. 2020). The dopamine was detected within a wide range of concentrations, from 0.3 to 750 μM, with the LOD of 0.2 μM.

DNA damage was successfully monitored using blue luminescence GQDs (Lu, Wei et al. 2014). It was observed that the GQDs showed the excitation-dependent PL and stable electrochemiluminescence (ECL) performances. This ECL sensor was able to detect p53 ssDNA, with the LOD of 13 nM.

Carcino-embryonic antigen (CEA) was successfully detected using an immunosensor based on N-doped GQDs (Yang, Liu et al. 2017). The N-doped GQDs were placed on PtPd bimetallic nanoparticles and this hybrid was then functionalized with Au nanoparticles. This hybrid material showed an outstanding electrocatalytic activity for the H_2O_2 reduction. The material was selected as transducing component with a role to capture antibodies. In this way the increase of the electrochemical response intensity was enhanced as well. A linear response was shown to manifestate itself in the concentration range from 5 fg/mL to 50 ng/mL with the LOD of 2 fg/mL. One more study showed the possibility of using GQDs in a ECL immunosensor for the CEA detection (Nie, Wang et al. 2018). As a signal amplifier the poly(5-formylindole)/reduced graphene oxide nanocomposite (P5FIn/erGO) was used. Herein, the broad linear concentration range from 0.1 pg/mL to 10 ng/mL with the low LOD of 3.78 fg/mL were reported.

Peanut allergen Ara h 1 is one more analyte which the GQDs sensor was able to detect (Li, Liu et al. 2018). Li et al. reported production of the gold-palladium nanowaxberries (AuPdNWs)/dodecylamine-functionalized graphene quantum dots (D-GQDs)-graphene micro-aerogel (GMA) composite. The sensor was constructed in the following manner: conductor, semiconductor, and conductor. For the detection of peanut allergen Ara h 1, a differential pulse voltammetric (DPV) response was monitored. The linear concentration range from 1.0×10^{-22} to 1.0×10^{-17} M was observed.

Cardiac troponin I (cTnI) was also detected with a GQD sensor (Fan, Bao et al. 2018). The PEC sensor with N,S-co-doped GQDs and CdS co-sensitized hierarchical Zn_2SnO_4 cube structure was constructed and explored for the cTnI detection by employing the specific immune recognition between the cTnI antigens and cTnI antibodies (anti-cTnI). This reaction lowered the intensity of the photoelectric signal and showed the linear response in the concentration range from 0.001 ng/mL to 50 ng/mL with the LOD of 0.3 pg/mL.

One more sensor for the detection of H_2O_2 in the clinical cancer tissue was developed using GQDs (Xu, Yuan et al. 2018). The electrode was modified by depositing AuPd alloy nanoparticles (AuPd-ANPs) together with GQDs, and the electrochemical sensing for the detection of cancer biomarker H_2O_2 in human breast cancer cells and tissue was studied. The linear concentration range was reported between 1.0 μM and 18.44 mM, with the LOD of 500 nM.

Norepinephrine, a stress hormone and neurotransmitter, was detected using an electrochemical sensor based on GQDs (Fajardo, Tapia et al. 2019). The linear concentration range was reported from 0.5 to 7.5 μmol/L, with the LOD of 0.15 μmol/L.

Hemoglobin was detected using a GQD-based sensor as well (Tran and Doong 2019). S-doped GQDs produced in the hydrothermal reactor were able to establish π-π interactions with hemoglobin, leading to the quenching of photoluminescence. The linear concentration range from 10 to 1000 nM was reported along with the low LOD of 0.48 nM for the analysis of the human serum.

Cholesterol in the human serum was one more analyte that a GQD sensor can detect (Abdolmohammad-Zadeh and Ahmadian 2021). The nanocomposite formed by GQDs with zirconium-based metal–organic framework showed a photocatalytic activity in peroxidase-like reactions. The mechanism of detection was based on the fluorescence quenching. Fabricated nanocomposite acts as a peroxidase where cholesterol was measured due to the terephthalic acid (TA) oxidation with H_2O_2 produced from the cholesterol oxidation in the presence of ChOx. The linear concentration range was reported to span from 0.04 to 1.60 μmol/L, with the LOD of 0.01 μmol/L.

Procalcitonin (PCT) is the precursor in the hormone calcitonin biosynthesis. It was successfully detected using an ECL immunosensor with GQDs (Hu, Song et al. 2021). Sandwich-type immunosensor components were isoreticular metal-organic framework-3 (IRMOF-3) enriched with the N-GQDs luminescent material (N-GQDs@IRMOF-3@N-GQDs, as donor) and Au NPs with zinc oxide nanorods (ZnO@Au, as acceptor). The range of detection was reported to be from 0.0001 to 100 ng/mL with the LOD of 12.58 fg/mL. The sensor was also investigated for the PCT detection in human serum and showed satisfying recovery.

2.4. Detection of Drugs Using GQDs Sensors

Diverse drugs, such as antibiotics, antiviral drugs, and proton pump inhibitors, were detected using GQDs sensors and in the following section some of them are discussed.

Rabeprazole, a drug that belongs to the proton pump inhibitors and helps in the regulation of stomach acid, was successfully detected using a chemiluminescence sensor containing GQDs (Yousefzadeh, Abolhasani et al. 2019). Silver nanoclusters (AgNCs) and GQDs were incorporated into the sensor and showed a synergetic effect by increasing the emission. The concentration of rabeprazole was linear in the range 4—133 ng/mL with the LOD of 1.1 ng/mL.

Ibuprofen detection was also achieved using an electrochemical sensor containing GQDs (Roushani and Shahdost-fard 2019). A glass carbon electrode (GCE) was modified with Ag nanoparticles and N-doped GQDs and the AuNPs@N-GQDs/GCE system was produced. An aptamer (Apt) molecule for the biological recognition of Ibuprofen (IBP) was loaded on the AuNPs@N-GQDs/GCE. The LOD in the range of aM, 33.33 aM, was reported, the lowest Ibuprofen concentration found.

Antibiotic sulfadiazine is another analyte detected in the concentration range 0.04–22.0 μM with the LOD of 10 nM using the GQDs/AgNPs composite as a fluorescent sensor (Afsharipour, Haji Shabani et al. 2020).

Antibiotic neomycin sulfate was quantitatively detected in the rubella vaccine (Pinto, Toloza et al. 2019). Amino-functionalized GQDs were used as a photoluminescent sensor with Fe^{3+} ions serving as a mediator. First, Fe^{3+} ions were added to the GQDs inducing the quenching of photoluminescence, then the neomycin sulfate was added, and PL was restored due to the ability of antibiotic to remove Fe^{3+} from the GQD surface. The recovery of the PL intensity was proportional to the neomycin sulfate concentration in the range from 10^{-7} to 10^{-5} mol/L and the method was successfully applied for the analysis of rubella vaccine for the presence of the neomycin trace residues.

Antiviral drug sofosbuvir (SOF) was detected using the electrochemical approach with molecularly imprinted polymers (MIP) forming composite with N,S-doped GQDs and gold nanoparticles (Mahmoud, El-Wekil et al. 2019). The extremely low LOD of 0.36 nM was obtained while the sensor showed the linear concentration response from 1 to 400 nM.

Sulfamethoxazole was also measured using a GQDs sensor (Le, Lee et al. 2020). A fluorescence sensor was produced using GQDs trapped inside a silica molecularly imprinted polymer (GQDs@SMIP). The schematic presentation of the structure of the material is shown in Figure 10.

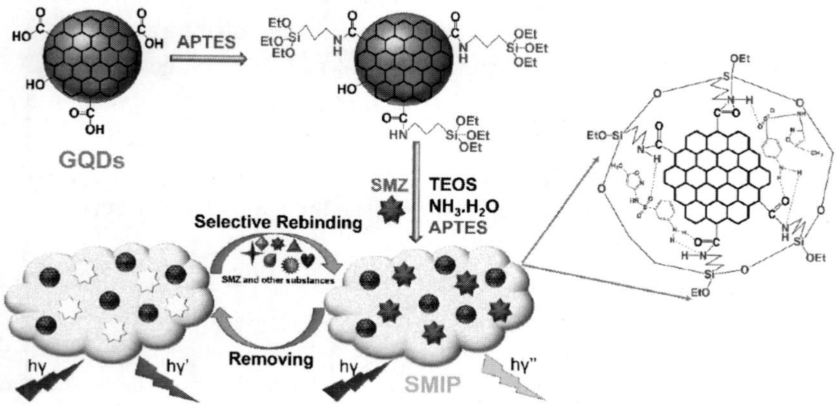

Figure 10. Schematic presentation of the GQDs@SMIP hybrid material production and the mechanism of detection. Figure adapted from Materials 2020, 13(11), 2521 (Le, Lee et al. 2020). Copyright 2020, MDPI AG.

The hybrid material GQDs@SMIP was able to absorb sulfamethoxazole which caused the quenching of the GQD photoluminescence. The PL intensity linear response was reported for concentrations of sulfamethoxazole in the range of 1 to 100 μM. This sensor was found to be selective for several compounds from the sulfonamide class (Figure 11).

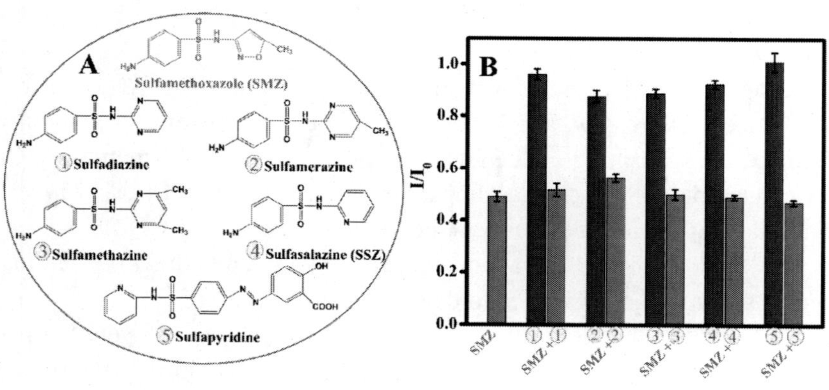

Figure 11. The structures of sulfonamide compounds (A) and their ability to quench photoluminescence of the GQDs@SMIP hybrid (B). Figure adapted from Materials 2020, 13(11), 2521 (Le, Lee et al. 2020). Copyright 2020, MDPI AG.

D-penicillamine (D-PA) was detected using a PL quenching sensor based on GQDs with the assistance of Pd^{2+} ions (Wang, Li et al. 2020). Herein, the GQDs were mixed with Pb^{2+} ions which led to the PL quenching due to the electron transfer. When D-PA was added, the PL intensity was recovered, as a result of Pb^{2+} removal from the dots surface. It was suggested that D-PA creates strong coordination bonds with the metal ions. The linear response was shown for the concentration range from 0.6 to 50 μmol/L of D-PA with the LOD of 0.47 μmol/L. The researchers also investigated the possibility of the D-PA detection in the real sample of humane urea and obtained recoveries of 96.84%–102.13%.

Small peptide glutathione (GSH) was also detected with the GQDs-based sensor (Li, Zhang et al. 2020). The GQDs were functionalized with amino groups, and Cu^{2+} ions were added inducing the PL quenching. When the analyte GSH was added, the PL was recovered due to the ion dissociation from the dots and the PL intensity was proportional to the GSH concentration. Herein, the linear concentration range was reported for 20–500 μmol/L with the LOD of 3.4 μmol/L.

Antibiotic tetracyclines (TCs) were measured in cow milk with a GQDs-based sensor as well (Ahmed, Kumar et al. 2021). The detection was based on the aptamer-induced fluorescence fluctuation of the GQDs and palladium nanoparticles (PdNPs) composite. The LOD of 45 ng/mL was achieved. The PdNPs were conjugated with aptamer, and then the GQDs were added and connected with aptamer by the π-π stacking. In this complex, the PL intensity of the GQDs was significantly lowered. By adding TCs, the GQDs were released from the complex and, consequently, the higher PL intensity was recorded.

Drug tiopronin together with the Cu^{2+} ions were detected using a fast GQD-based sensor (Yao and Wang 2021). The mechanism of detection was based on the quenching of the GQDs fluorescence with the increasing Cu^{2+} ion concentration, which was linear in the range 0.01-0.5 μM. Then, the addition of a drug in the system caused the recovery of the GQD fluorescence, due to the stronger interaction between ions and drugs. The linear concentration range was found to be from 0.01 to 0.4 μM.

Antibiotic gentamicin was detected with a GQDs-AuNPs sensor in cow milk and eggs (Sajwan and Solanki 2022). The mechanism of detection was the "on-off" fluorescence. The linear range was achieved for the concentrations from 1.03 to 16.55 μM, and the LOD was reported to be of 0.493 μM. The sensor also showed a linear response for colorimetric measurements with the LOD of 0.422 μM.

2.5. Detection of Pathogen Microorganisms (Viruses, Bacteria, and Cancer Cells) with GQDs

Detection of the hepatitis C virus core antigen (HCV) was achieved using a GQDs-based biosensor (Valipour and Roushani 2017). Thiol functionalized GQDs were immobilized with AgNPs and used as a support for the HCV antibody and loaded on the electrode surface. The lowering of the electrochemical signal due to the specific antibodies—antigens recognition was observed. The very broad linear concentration range was achieved (from 0.05 pg/mL to 60 ng/mL) with the LOD of 3 fg/mL. The produced immunosensor was suitable for the analysis of the HCV antigen in the serum.

Tumor suppressor protein p53 was detected using GQDs/gold nanoparticles as dual amplification components of an immunosensor (Hasanzadeh, Baghban et al. 2018). The linear concentration range was reported from 0.000197 to 0.016 pM with the LOD of 0.065 fM.

Adenovirus was also detected using a GQDs sensor (Ahmed, Mogus et al. 2018). This optoelectronic sensor was able to detect adenoviruses with the sensitivity of 8.75 PFU/mL which is 100 times more sensitive than the conventional enzyme-linked immunosorbent assay (ELISA) method. The carbon electrode for sensing was covered with nanobundles made of gold (AuNBs). Separately, GQDs were produced and virus antibodies were conjugated with both nanomaterials. By adding the adenovirus, the antibody-antigen interaction was established causing the lowering of the distance between the AuNB and GQDs, and the light-induced interaction produced a local electric signal. This sensor was able to detect the virus at the very low concentrations in complex biological samples.

A similar system was proposed for the detection of the hepatitis B surface antigen (Yan, Yang et al. 2018). N-doped GQDs were supported by multi-metal nanoparticles, AuPdCu. The resulting complex AuPdCu/N-GQDs showed electrical conductivity and outstanding catalytic activity for the H_2O_2 reduction. Then as a carrier and electronic mediator to load complex nanomaterial the polymer nanospheres were used. The polymer improved capacity for the antibodies loading and enhanced the electron transport velocity. This immunosensor showed a very good sensitivity with the linear concentration range from 10 fg/mL to 50 ng/mL. The LOD was reported to be 3.3 fg/mL.

Von Willebrand factor (vWF) was detected using GQDs-based immunosensor (Kong, Li et al. 2018). First, vWF antibodies were bound to Ag@AuNPs, and the Ag@Au-Ab nanocomposites were produced. With the

GQDs, this complex quenched PL. But when the vWF and Ag@Au-Ab NCs reacted, the PL of GQDs was recovered. The linear concentration range was from 0.1 pg/mL to 10 ng/mL with the LOD of 30 fg/mL.

Mycobacterium tuberculosis antigen (culture filtrate protein; CFP-10) was detected using an electrochemical biosensor with GQDs (Tufa, Oh et al. 2018). Herein, a glassy carbon electrode (GCE) was covered with the complex nanomaterial, GQD-coated Fe_3O_4@Ag core-shell nanostructure (Fe_3O_4@Ag/GQD). The broad linear range for the detection of pathogenic bacteria was achieved, from 0.005 to 500 μg/mL, with the LOD of 0.33 ng/mL.

One more pathogen, Leptospira interrogans, was detected using an electrochemical sensor based on GQDs (Nagraik, Kaushal et al. 2020). The sensing electrode was produced from carboxylated multiwalled carbon nanotubes with the AuNPs (c-MWCNTs/nanoAu) electrode and GQDs that were bound to the electrode using 4-aminothiophenol. The LipL32 gene of Leptospira interrogans was detected by a change in an electrochemical signal using differential pulse voltammetry (DPV) and methylene blue (MB). Molecules of MB interacted with single strand DNA, accumulated and increased the conductivity. The linear concentration range was from 0.37 to 12 ng/μl.

White spot syndrome virus (WSSV) was detected using a disposable electrode produced from conductive silicon rubber, S,N-doped GQDs, and AuNPs-polyaniline nanocomposite, and the anti-WSSVs antibody was bonded on the composite surface (Takemura, Satoh et al. 2020). The impedimetric sensing results were found to be linear from 1.45×10^2 to 1.45×10^5 DNA copies/mL, with the LOD of 48.4 DNA copies/mL.

Glioma is the predominant brain tumor and the detection of glioma cells at low concentration is highly important and lifesaving. A sensor for detection of these cells was produced using GQDs (Ganganboina, Dega et al. 2021). This sensor was an impedimetric biosensor produced using the S-doped GQDs placed on the surface of AuNPs surrounded by carbon nanospheres (Au–CNS). This S-GQDs@Au–CNS nanocomposite has two functions, one to increase the electrochemical activity and another to ensure the selectivity through the angiopep-2 (Ang-2) specific for glioma cells. The very broad cell concentration range was achieved, from 100 to 100,000 cells/mL, with the LOD of 40 cells/mL. The biosensor was suitable for the cell detection in the human serum.

Breast cancer cells were also detected with an electrochemical biosensor based on GQDs (Pothipor, Jakmunee et al. 2021). Biomarkers miRNA-21, miRNA-155, and miRNA-210 were detected using the AuNPs/GQDs/

graphene oxide modified carbon electrodes. The linear concentration range was achieved from 0.001 to 1000 pM with the LODs of 0.04, 0.33, and 0.28 fM for miRNA-21, miRNA-155, and miRNA-210, respectively.

2.6. Pesticide Detection with GQDs

Detection of pesticides in food and the environment is a very important aspect in control of the pesticide use, health, and nature protection. Thus, sensors for this purpose are also needed and there are only several papers that investigate the application of GQDs in this area.

Pesticide tributyltin, a biocide, was detected using the magnetic silica beads/GQDs/molecularly imprinted polypyrrole (mSGP) composite (Zor, Morales-Narváez et al. 2015). The polymer was tailored to capture small molecules and preconcentrate them on the sensor surface, and the PL of the GQDs was lowered when the analyte was recognized. The sensor showed high selectivity for the detection of tributyltin directly in seawater, without any sample treatment.

Paraoxon was detected using the fluorescence quenching mechanism of GQDs with acetylcholinesterase (Li, Wang et al. 2015).

Pentachlorophenol (PCP) was detected in the concentration range from 0.02 to 300 ng/mL using the Cu_2O/GQDs composite in the sensor in ECL (Yan, Liu et al. 2016).

Chlorpyrifos was detected in the concentration ranges from 3.2×10^{-8} to 6×10^{-7} M (Mao, Zuo et al. 2017) and 1×10^{-16}—1.0×10^{-11} M with the detection limit 3.8×10^{-17} M (Ruiyi, Hui et al. 2022), and 2-methyl-4-chlorophenoxyacetic acid within the concentration range from 10 pmol/L to 0.1 μmol/L (Yang, Fang et al. 2017) was detected with the electrochemical GQDs sensors. N-doped GQDs were coupled with omethoate aptamer, and the LOD was reportec to be of 0.041 nM (Zhang, Lin et al. 2017), while later the S-doped GQDs were used as well with the LOD of 4.7×10^{-9} M (or 0.001 ppm) (Nair, Chandran et al. 2021). Parathion was detected in the concentration range 0.01–106 ng/L using GQDs in a screen-printed electrochemical immunosensor (Mehta, Bhardwaj et al. 2017). Dichlorvos was detected by combining GQDs, acetylcholinesterase (AChE), and choline oxidase (CHOx) (Sahub, Tuntulani et al. 2018), where the LOD was reported to be 0.172 ppm (0.778 μM). Detection was based on the fluorescence "off" mechanism. Thiacloprid (Liu, Cao et al. 2018) and atrazine (Wang, Wang et al. 2019) were detected too. Glyphosat was detected using the GQDs-AgNPs composite and the PL quenching was linear relative to pesticide concentration (Jiménez-López, Llorent-Martínez et al. 2020).

Both the pesticide Malathion and Hg^{2+} ions were detected using GQDs as a sensing agent (Roushani, Kohzadi et al. 2018). Herein, negatively charged carboxyl groups of GQDs and Hg^{2+} were bound coordinately which resulted in the quenching of the GQDs fluorescence. When Malathion was added, the GQD fluorescence was restored. In this phase, the interaction between Hg^{2+} and Malathion was described as a strong metal-thiol bond. A linear dependence between the Hg^{2+} concentration and PL intensity was shown in two concentration ranges of 1–10 μM and 10–80, while for Malathion the linear concentration range was reported from 80 to 120 μM, with the LOD of 0.5 μM.

Carbamate, an insecticide that kills insects causing reversibly inactivation of acetylcholinesterase, was detected with S-doped GQDs (Nair, Thomas et al. 2020). The dots were incorporated in the polymer poly(vinyl)alcohol (PVA) and a PVA/S-GQD film was produced. Pesticides carbofuran and thiram were detected due to their ability to quench the PL intensity of the S-GQDs by the photoinduced electron transfer. It was suggested that O and S atoms in the GQDs structures due to their higher electronegativity form hydrogen bonds with carbamates. The linear ratio between the pesticide concentration and the PL signal was established in the concentration range 0–500 ppb with the LOD of 2.03×10^{-12} M for carbofuran.

Pesticide carbendazim was detected in strawberries using an arginine-functionalized ruthenium-GQDs hybrid (Xu, Li et al. 2021). The differential pulse voltammetry method detected a linear response when the carbendazim concentration changed from 0.01 μM to 45 μM. The LOD of 0.004 μM was reported.

Trifluralin pesticide was detected with the LOD of 10 nM (Mahmoudpour, Saadati et al. 2021) when GQDs were placed together with AuNPs and AgNPs on the surface of the robber glove.

Conclusion

Graphene quantum dots have been inspiring scientists from different fields to create novel, astonishing, groundbreaking materials and devices. Due to their outstanding properties, such as biocompatibility, photoluminescence, chemical stability, and various ways for modification, GQDs have easily created their path to become a part of components used in the everyday life.

One of the most important and broadly studied applications of GQDs is the detection of different analytes. The analytes can be divided into the following groups:

1. ions
2. organic compounds
3. metabolites
4. pathogenic microorganisms.

Different types of sensors based on GQDs have been developed. So far, a large number of ions were detected with GQD sensors: Hg^{2+}, Co^{2+}, Pb^{2+}, Fe^{3+}, Ca^{2+}, U^{6+}, As^{3+}, Ag^{+}, Al^{3+}, Au^{3+}, CN^{-}, Cd^{2+}, Mn^{2+}, S^{2-}, Cr^{3+}, Ru^{3+}, Ni^{2+}, and others. The methods for detection of drugs are mainly TLC, GC, and HPLC. These instruments are expensive, demands a time, professional staff, complex sample preparation. Due to the size of instruments, these techniques are not transferable to portable technologies. Thus, development of GQDs sensors as a cost-effective, fast, portable, eco-friendly, simple tool will largely shortened testing time and lowered price of analysis. So far, GDQs have been investigated for the detection of the following drugs: rabeprazole, ibuprofen, sulfadiazine, neomycin, sulfate, sofosbuvir, sulfamethoxazole, sulfamethoxazole, D-penicillamine, glutathione, tetracyclines, tiopronin, gentamicin, amphetamine, and cocaine. Various metabolites have been detected with the GQDs sensors as well: procalcitonin, cholesterol, hemoglobin, norepinephrine, H_2O_2, cardiac troponin I, peanut allergen Ara h 1, arcino-embryonic antigen, dopamine, prostate-specific antigen, glucose etc.

Detection of microorganisms and pathogens is particularly important and the application of GQDs has significantly improved their detection. GQDs are in most cases a component of electrodes for electrochemical and electroluminescence detection. The following pathogens have been detected so far: hepatitis C virus core antigen, tumor suppressor protein p53, adenovirus, hepatitis B surface antigen, Von Willebrand factor, mycobacterium tuberculosis antigen, Bacillus anthracis spores, Leptospira interrogans, white spot syndrome virus, glioma cells, breast cancer cells. Although GQDs have been studied for the detection of just a few pesticides for the present time, the authors of this chapter believe that this field will be highly explored in the future, particularly the fluorescence quenching detection, due to low price, simplicity, speed of sample preparation, and measurements.

Acknowledgment

The research was supported by Science Fond of the Republic of Serbia, 7741955 GRANT No, "Are photoactive nanoparticles salvation for global infectonal treath?" - PHOTOGUN4MICROBES and by the Ministry of Education, Science and Technological Development of the Republic of Serbia [grant number 451-03-68/2022-14/200017].

Dedication

To generations that are coming, to Lenka and Milan for a better future....

References

Abbas, A., T. A. Tabish, S. J. Bull, T. M. Lim and A. N. Phan (2020). "High yield synthesis of graphene quantum dots from biomass waste as a highly selective probe for Fe3+ sensing." *Scientific Reports* 10(1): 21262.

Abdolmohammad-Zadeh, H., and F. Ahmadian (2021). "A fluorescent biosensor based on graphene quantum dots/zirconium-based metal-organic framework nanocomposite as a peroxidase mimic for cholesterol monitoring in human serum." *Microchemical Journal* 164: 106001.

Abolghasemi-Fakhri, Z., and M. Amjadi (2021). "Gold nanostar@graphene quantum dot as a new colorimetric sensing platform for detection of cysteine." *Spectrochimica Acta - Part A: Molecular and Biomolecular Spectroscopy* 261: 120010.

Achadu, O. J., and T. Nyokong (2017). "Graphene quantum dots anchored onto mercaptopyridine-substituted zinc phthalocyanine-Au@Ag nanoparticle hybrid: Application as fluorescence "off-on-off" sensor for Hg^{2+} and biothiols." *Dyes and Pigments* 145: 189-201.

Afsharipour, R., A. M. Haji Shabani, S. Dadfarnia and E. Kazemi (2020). "Selective fluorometric determination of sulfadiazine based on the growth of silver nanoparticles on graphene quantum dots." *Microchimica Acta* 187(1): 54.

Ahmadi, N., M. Bagherzadeh and A. Nemati (2020). "Comparison between electrochemical and photoelectrochemical detection of dopamine based on titania-ceria-graphene quantum dots nanocomposite." *Biosensors and Bioelectronics* 151: 111977.

Ahmed, S. R., J. Mogus, R. Chand, E. Nagy and S. Neethirajan (2018). "Optoelectronic fowl adenovirus detection based on local electric field enhancement on graphene quantum dots and gold nanobundle hybrid." *Biosensors and Bioelectronics* 103: 45-53.

Ahmed, S. R., S. Kumar, G. A. Ortega, S. Srinivasan and A. R. Rajabzadeh (2021). "Target specific aptamer-induced self-assembly of fluorescent graphene quantum dots on palladium nanoparticles for sensitive detection of tetracycline in raw milk." *Food Chemistry* 346: 128893.

Ahmed, S. Rahin, M. Sherazee, S. Srinivasan and A. Reza Rajabzadeh (2022). "Nanozymatic detection of thiocyanate through accelerating the growth of ultra-small gold nanoparticles/graphene quantum dots hybrids." *Food Chemistry* 379: 132152.

Amjadi, M., R. Shokri and T. Hallaj (2016). "A new turn-off fluorescence probe based on graphene quantum dots for detection of Au(III) ion." *Spectrochimica Acta - Part A: Molecular and Biomolecular Spectroscopy* 153: 619-624.

Anas, N. A. A., Y. W. Fen, N. A. S. Omar, W. M. E. M. M. Daniyal, N. S. M. Ramdzan and S. Saleviter (2019). "Development of Graphene Quantum Dots-Based Optical Sensor for Toxic Metal Ion Detection." *Sensors* 19(18): 3850.

Anh, N. T. N., A. D. Chowdhury and R. A. Doong (2017). "Highly sensitive and selective detection of mercury ions using N, S-codoped graphene quantum dots and its paper strip based sensing application in wastewater." *Sensors and Actuators, B: Chemical* 252: 1169-1178.

Anh, N. T. N., and R. A. Doong (2018). "One-Step Synthesis of Size-Tunable Gold@Sulfur-Doped Graphene Quantum Dot Nanocomposites for Highly Selective and Sensitive Detection of Nanomolar 4-Nitrophenol in Aqueous Solutions with Complex Matrix." *ACS Applied Nano Materials* 1(5): 2153-2163.

Arnault, J. C., (2021). Chapter 4 - Nanodiamonds: From synthesis to applications. *Handbook of Carbon-Based Nanomaterials*. S. Thomas, C. Sarathchandran, S. A. Ilangovan and J. C. Moreno-Piraján, Elsevier: 209-246.

Ashraf, G., M. Asif, A. Aziz, A. Q. Dao, T. Zhang, T. Iftikhar, Q. Wang and H. Liu (2020). "Facet-energy inspired metal oxide extended hexapods decorated with graphene quantum dots: Sensitive detection of bisphenol A in live cells." *Nanoscale* 12(16): 9014-9023.

Barone, P. W., S. Baik, D. A. Heller and M. S. Strano (2005). "Near-infrared optical sensors based on single-walled carbon nanotubes." *Nature Materials* 4(1): 86-92.

Bhardwaj, H., M. K. Pandey, Rajesh and G. Sumana (2019). "Electrochemical Aflatoxin B1 immunosensor based on the use of graphene quantum dots and gold nanoparticles." *Microchimica Acta* 186(8): 592.

Biswas, M. C., M. T. Islam, P. K. Nandy and M. M. Hossain (2021). "Graphene Quantum Dots (GQDs) for Bioimaging and Drug Delivery Applications: A Review." *ACS Materials Letters* 3(6): 889-911.

Bressi, V. A., Ferlazzo, D. Iannazzo and C. Espro (2021). "Graphene Quantum Dots by Eco-Friendly Green Synthesis for Electrochemical Sensing: Recent Advances and Future Perspectives." *Nanomaterials* 11(5): 1120.

Cha, C., S. R. Shin, N. Annabi, M. R. Dokmeci and A. Khademhosseini (2013). "Carbon-based nanomaterials: multifunctional materials for biomedical engineering." *ACS Nano* 7(4): 2891-2897.

Chen, C., D. Zhao, T. Hu, J. Sun and X. Yang (2017). "Highly fluorescent nitrogen and sulfur co-doped graphene quantum dots for an inner filter effect-based cyanide sensor." *Sensors and Actuators, B: Chemical* 241: 779-788.

Chen, J., W. Wu, F. Zhang, J. Zhang, H. Liu, J. Zheng, S. Guo and J. Zhang (2020). "Graphene quantum dots in photodynamic therapy." *Nanoscale Advances* 2(10): 4961-4967.

Chen, L., C. Wu, P. Du, X. Feng, P. Wu and C. Cai (2017). "Electrolyzing synthesis of boron-doped graphene quantum dots for fluorescence determination of Fe^{3+} ions in water samples." *Talanta* 164: 100-109.

Chen, S. X. Hai, X. W. Chen and J. H. Wang (2014). "In Situ growth of silver nanoparticles on graphene quantum dots for ultrasensitive colorimetric detection of H_2O_2 and glucose." *Analytical Chemistry* 86(13): 6689-6694.

Chen, W., R. Li, Z. Li, Y. Yang, H. Zhu and J. Liu (2019). "Promising copper oxide-histidine functionalized graphene quantum dots hybrid for electrochemical detection of hydroquinone." *Journal of Alloys and Compounds* 777: 1001-1009.

Chen, Y. X., D. Lu, G.-G. Wang, J. Huangfu, Q.-B. Wu, X.-F. Wang, L.-F. Liu, D.-M. Ye, B. Yan and J. Han (2020). "Highly Efficient Orange Emissive Graphene Quantum Dots Prepared by Acid-Free Method for White LEDs." *ACS Sustainable Chemistry & Engineering* 8(17): 6657-6666.

Chen, Z., Z. Zhao, J. Liu, X. Zhu, L. Wang, Y. Xu, H. Xu and H. Li (2018). "Graphene quantum dots modified Ag3PO4 for facile synthesis and the enhanced photocatalytic performance." *Journal of the Chinese Advanced Materials Society* 6(3): 255-269.

Chirea, M. (2013). "Electron Transfer at Gold Nanostar Assemblies: A Study of Shape Stability and Surface Density Influence." *Catalysts* 3(1): 288-309.

Chung, S., R. A. Revia and M. Zhang (2021). "Graphene Quantum Dots and Their Applications in Bioimaging, Biosensing, and Therapy." *Advanced Materials* 33(22): 1904362.

Ciotta, E., S. Paoloni, M. Richetta, P. Prosposito, P. Tagliatesta, C. Lorecchio, I. Venditti, I. Fratoddi, S. Casciardi and R. Pizzoferrato (2017). "Sensitivity to Heavy-Metal Ions of Unfolded Fullerene Quantum Dots." *Sensors* 17(11): 2614.

Das, R., K. K. Paul and P. K. Giri (2019). "Highly sensitive and selective label-free detection of dopamine in human serum based on nitrogen-doped graphene quantum dots decorated on Au nanoparticles: Mechanistic insights through microscopic and spectroscopic studies." *Applied Surface Science* 490: 318-330.

De Acha, N., C. Elosua, I. Matias and F. J. Arregui (2017). "Luminescence-Based Optical Sensors Fabricated by Means of the Layer-by-Layer Nano-Assembly Technique." *Sensors* (Basel, Switzerland) 17(12): 2826.

Dewangan, P. K., F. Khan, K. Shrivas and V. Sahu (2019). "Determination of uranium in environmental sample by nanosensor graphene quantum dots." *Journal of Radioanalytical and Nuclear Chemistry* 320(3): 757-763.

Dorontić, S., S. Jovanović and A. Bonasera (2021). "Shedding Light on Graphene Quantum Dots: Key Synthetic Strategies, Characterization Tools, and Cutting-Edge Applications." *Materials* 14(20): 6153.

Du, D., K. Wang, Y. Wen, Y. Li and Y. Y. Li (2016). "Photodynamic Graphene Quantum Dot: Reduction Condition Regulated Photoactivity and Size Dependent Efficacy." *ACS Appl Mater Interfaces* 8(5): 3287-3294.

Facure, M. H. M., R. Schneider, J. B. S. Lima, L. A. Mercante and D. S. Correa (2021). "Graphene Quantum Dots-Based Nanocomposites Applied in Electrochemical Sensors: A Recent Survey." *Electrochem* 2(3): 490-519.

Fajardo, A., D. Tapia, J. Pizarro, R. Segura and P. Jara (2019). "Determination of norepinephrine using a glassy carbon electrode modified with graphene quantum dots and gold nanoparticles by square wave stripping voltammetry." *Journal of Applied Electrochemistry* 49(4): 423-432.

Fan, D., C. Bao, M. S. Khan, C. Wang, Y. Zhang, Q. Liu, X. Zhang and Q. Wei (2018). "A novel label-free photoelectrochemical sensor based on N,S-GQDs and CdS co-sensitized hierarchical Zn_2SnO_4 cube for detection of cardiac troponin I." *Biosensors and Bioelectronics* 106: 14-20.

Faridbod, F., V. K. Gupta and H. A. Zamani (2011). "Electrochemical Sensors and Biosensors." *International Journal of Electrochemistry* 2011: 352546.

Farmani, M. R., H. Peyman and H. Roshanfekr (2020). "Blue luminescent graphene quantum dot conjugated cysteamine functionalized-gold nanoparticles (GQD-AuNPs) for sensing hazardous dye Erythrosine B." *Spectrochimica Acta - Part A: Molecular and Biomolecular Spectroscopy* 229: 117960.

Feng, Y., J. Zhao, X. Yan, F. Tang and Q. Xue (2014). "Enhancement in the fluorescence of graphene quantum dots by hydrazine hydrate reduction." *Carbon* 66: 334–339.

Fu, C. C., C. T. Hsieh, R. S. Juang, S. Gu, Y. Ashraf Gandomi, R. E. Kelly and K. D. Kihm (2020). "Electrochemical sensing of mercury ions in electrolyte solutions by nitrogen-doped graphene quantum dot electrodes at ultralow concentrations." *Journal of Molecular Liquids* 302: 112593.

Ganganboina, A. B., N. K. Dega, H. L. Tran, W. Darmonto and R. A. Doong (2021). "Application of sulfur-doped graphene quantum dots@gold-carbon nanosphere for electrical pulse-induced impedimetric detection of glioma cells." *Biosensors and Bioelectronics* 181: 113151.

Gao, B., D. Chen, B. Gu, T. Wang, Z. Wang, F. xie, Y. Yang, Q. Guo and G. Wang (2020). "Facile and highly effective synthesis of nitrogen-doped graphene quantum dots as a fluorescent sensing probe for Cu^{2+} detection." *Current Applied Physics* 20(4): 538-544.

Gao, X. X., X. Zhou, Y. F. Ma, C. P. Wang and F. X. Chu (2018). "A fluorometric and colorimetric dual-mode sensor based on nitrogen and iron co-doped graphene quantum dots for detection of ferric ions in biological fluids and cellular imaging." *New Journal of Chemistry* 42(18): 14751-14756.

Geim, A. K., and K. S. Novoselov (2007). "The rise of graphene." *Nature Materials* 6(3): 183-191.

Gevaerd, A., C. E. Banks, M. F. Bergamini and L. H. Marcolino-Junior (2020). "Nanomodified Screen-Printed Electrode for direct determination of Aflatoxin B1 in malted barley samples." *Sensors and Actuators B: Chemical* 307: 127547.

Ghalkhani, M., E. M. Khosrowshahi and E. Sohouli (2021). Chapter 3 - Carbon nano-onions: Synthesis, characterization, and application. *Handbook of Carbon-Based Nanomaterials.* S. Thomas, C. Sarathchandran, S. A. Ilangovan and J. C. Moreno-Piraján, Elsevier: 159-207.

Gu, S., C. T. Hsieh, Y. Y. Tsai, Y. Ashraf Gandomi, S. Yeom, K. D. Kihm, C. C. Fu and R. S. Juang (2019). "Sulfur and Nitrogen Co-Doped Graphene Quantum Dots as a Fluorescent Quenching Probe for Highly Sensitive Detection toward Mercury Ions." *ACS Applied Nano Materials* 2(2): 790-798.

Guin, S. K., A. S. Ambolikar, J. Paul Guin and S. Neogy (2018). "Exploring the excellent photophysical and electrochemical properties of graphene quantum dots for complementary sensing of uranium." *Sensors and Actuators, B: Chemical* 272: 559-573.

Hai, X., X. Lin, X. Chen and J. Wang (2018). "Highly selective and sensitive detection of cysteine with a graphene quantum dots-gold nanoparticles based core-shell nanosensor." *Sensors and Actuators, B: Chemical* 257: 228-236.

Hasanzadeh, M., H. N. Baghban, N. Shadjou and A. Mokhtarzadeh (2018). "Ultrasensitive electrochemical immunosensing of tumor suppressor protein p53 in unprocessed human plasma and cell lysates using a novel nanocomposite based on poly-cysteine/graphene quantum dots/gold nanoparticle." *International Journal of Biological Macromolecules* 107 (PartA): 1348-1363.

He, Y., C. Hu, Z. Li, C. Wu, Y. Zeng and C. Peng (2022). "Multifunctional carbon nanomaterials for diagnostic applications in infectious diseases and tumors." *Materials Today Bio* 14: 100231.

Henna, T. K., and K. Pramod (2020). "Graphene quantum dots redefine nanobiomedicine." *Materials Science and Engineering C* 110: 110651.

Hou, J., F. Bei, M. Wang and S. Ai (2013). "Electrochemical determination of malachite green at graphene quantum dots-gold nanoparticles multilayers-modified glassy carbon electrode." *Journal of Applied Electrochemistry* 43(7): 689-696.

Hu, L., C. Song, T. Shi, Q. Cui, L. Yang, X. Li, D. Wu, H. Ma, Y. Zhang, Q. Wei and H. Ju (2021). "Dual-quenching electrochemiluminescence resonance energy transfer system from IRMOF-3 coreaction accelerator enriched nitrogen-doped GQDs to ZnO@Au for sensitive detection of procalcitonin." *Sensors and Actuators B: Chemical* 346: 130495.

Huang, H., L. Liao, X. Xu, M. Zou, F. Liu and N. Li (2013). "The electron-transfer based interaction between transition metal ions and photoluminescent graphene quantum dots (GQDs): A platform for metal ion sensing." *Talanta* 117: 152-157.

Huang, Y., W. Zhou, Y. Wang and Y. Zhang (2020). "Crown ether-like structure in graphene quantum dots: Ultra-sensitive photoluminescence sensor for Ca^{2+} in vitro." *Synthetic Metals* 270: 116581.

Iannazzo, D., A. Pistone, M. Salamò, S. Galvagno, R. Romeo, S. V. Giofré, C. Branca, G. Visalli and A. Di Pietro (2017). "Graphene quantum dots for cancer targeted drug delivery." *International Journal of Pharmaceutics* 518(1-2): 185-192.

Iqbal, S., H. Khatoon, A. Hussain Pandit and S. Ahmad (2019). "Recent development of carbon based materials for energy storage devices." *Materials Science for Energy Technologies* 2(3): 417-428.

Jiménez-López, J., E. J. Llorent-Martínez, P. Ortega-Barrales and A. Ruiz-Medina (2020). "Graphene quantum dots-silver nanoparticles as a novel sensitive and selective luminescence probe for the detection of glyphosate in food samples." *Talanta* 207: 120344.

Jin, S. H., D. H. Kim, G. H. Jun, S. H. Hong and S. Jeon (2013). "Tuning the Photoluminescence of Graphene Quantum Dots through the Charge Transfer Effect of Functional Groups." *ACS Nano* 7(2): 1239-1245.

Jin, X., C. Feng, D. Ponnamma, Z. Yi, J. Parameswaranpillai, S. Thomas and N. V. Salim (2020). "Review on exploration of graphene in the design and engineering of smart sensors, actuators and soft robotics." *Chemical Engineering Journal Advances* 4: 100034.

Jovanović, S. P., Z. Syrgiannis, M. D. Budimir, D. D. Milivojević, D. J. Jovanovic, V. B. Pavlović, J. M. Papan, M. Bartenwerfer, M. M. Mojsin, M. J. Stevanović and B. M. Todorović Marković (2020). "Graphene quantum dots as singlet oxygen producer or radical quencher - The matter of functionalization with urea/thiourea." *Materials Science and Engineering: C* 109: 110539.

Jovanović, S., (2021). Gamma rays in carbon-based nanomaterials modification. Gamma Irradiation: Properties, Effects and Development of New Materials. *J. Brock.* Hauppauge, NY 11788 USA, Nova Science Publishers, Inc.: 65-121.

Jovanović, S., Z. Marković, M. Budimir, Z. Spitalsky, B. Vidoeski and B. Todorović Marković (2016). "Effects of low gamma irradiation dose on the photoluminescence properties of graphene quantum dots." *Optical and Quantum Electronics* 48(4): 259.

Jovanović, S., S. Dorontić, D. Jovanović, G. Ciasca, M. Budimir, A. Bonasera, M. Scopelliti, O. Marković and B. Todorović Marković (2020). "Gamma irradiation of graphene quantum dots with ethylenediamine: Antioxidant for ion sensing." *Ceramics International* 46(15): 23611-23622.

Jovanović, S., Z. M. Marković, Z. Syrgiannis, M. D. Dramićanin, F. Arcudi, V. L. Parola, M. D. Budimir and B. M. T. Marković (2017). "Enhancing photoluminescence of graphene quantum dots by thermal annealing of the graphite precursor." *Materials Research Bulletin* 93: 183-193.

Jovanović, S., Z. Marković and B. T. Marković (2017). Carbon based nanomaterials as agents for photodynamic therapy. *Photodynamic Therapy* (PDT): Principles, Mechanisms and Applications. F. Fitzgerald. New York, Nova Science Publishers: 35-108.

Jovanovic, S., Z. Syrgiannis, Z. M. Markovic, A. Bonasera, D. P. Kepic, M. D. Budimir, D. D. Milivojevic, V. D. Spasojevic, M. D. Dramicanin, V. B. Pavlovic and B. M. Todorovic Markovic (2015). "Modification of Structural and Luminescence Properties of Graphene Quantum Dots by Gamma Irradiation and Their Application in a Photodynamic Therapy." *ACS Appl Mater Interfaces* 7(46): 25865-25874.

Ju, J., and W. Chen (2014). "Synthesis of highly fluorescent nitrogen-doped graphene quantum dots for sensitive, label-free detection of Fe (III) in aqueous media." *Biosensors and Bioelectronics* 58: 219-225.

Kappen, J., S. Bharathi and S. A. John (2022). "Probing the Interaction of Heavy and Transition Metal Ions with Silver Nanoparticles Decorated on Graphene Quantum Dots by Spectroscopic and Microscopic Methods." *Langmuir,* 38, 4442–4451.

Keshipour, S., S. Nejati and M. Seyyedhamzeh (2021). "Graphene quantum dots modified with metformin/Co(II) as efficient oxidation catalyst" *Iranian Journal of Chemistry and Chemical Engineering (IJCCE),* 41, 3314.

Kong, L., Y. Li, C. Ma, B. Liu and L. Tan (2018). "Sensitive immunoassay of von Willebrand factor based on fluorescence resonance energy transfer between graphene quantum dots and Ag@Au nanoparticles." *Colloids and Surfaces B: Biointerfaces* 165: 286-292.

Kroto, H. W., J. R. Heath, S. C. O'Brien, R. F. Curl and R. E. Smalley (1985). "C_{60}: Buckminsterfullerene." *Nature* 318(6042): 162-163.

Kumar, Y. R., K. Deshmukh, K. K. Sadasivuni and S. K. K. Pasha (2020). "Graphene quantum dot based materials for sensing, bio-imaging and energy storage applications: a review." *RSC Advances* 10(40): 23861-23898.

Kuo, W. S., C. Y. Chang, H. H. Chen, C. L. Hsu, J. Y. Wang, H. F. Kao, L. C. Chou, Y. C. Chen, S. J. Chen, W. T. Chang, S. W. Tseng, P. C. Wu and Y. C. Pu (2016). "Two-Photon Photoexcited Photodynamic Therapy and Contrast Agent with Antimicrobial Graphene Quantum Dots." *ACS Appl Mater Interfaces* 8(44): 30467-30474.

Lai, S., Y. Jin, L. Shi, Y. Li and R. Zhou (2022). "Tailoring multifunctional gel for sensitive naked-eye and fluorescence dual-mode detection and effective adsorption of cadmium(II) and lead(II) ions in water." *Chemical Engineering Journal* 429: 132367.

Le, T. H., H. J. Lee, J. H. Kim and S. J. Park (2020). "Highly Selective Fluorescence Sensor Based on Graphene Quantum Dots for Sulfamethoxazole Determination." *Materials* 13(11): 2521.

Li, J., J. Qu, R. Yang, L. Qu and P. de B. Harrington (2016). "A Sensitive and Selective Electrochemical Sensor Based on Graphene Quantum Dot/Gold Nanoparticle Nanocomposite Modified Electrode for the Determination of Quercetin in Biological Samples." *Electroanalysis* 28(6): 1322-1330.

Li, K., W. Liu, Y. Ni, D. Li, D. Lin, Z. Su and G. Wei (2017). "Technical synthesis and biomedical applications of graphene quantum dots." *Journal of Material Chemistry B* 5(25): 4811-4826.

Li, L., G. Wu, T. Hong, Z. Yin, D. Sun, E. S. Abdel-Halim and J.-J. Zhu (2014). "Graphene Quantum Dots as Fluorescence Probes for Turn-off Sensing of Melamine in the Presence of Hg^{2+}." *ACS Applied Materials & Interfaces* 6(4): 2858-2864.

Li, L., L. Li, C. Wang, K. Liu, R. Zhu, H. Qiang and Y. Lin (2015). "Synthesis of nitrogen-doped and amino acid-functionalized graphene quantum dots from glycine, and their application to the fluorometric determination of ferric ion." *Microchimica Acta* 182(3-4): 763-770.

Li, N., X. Wang, J. Chen, L. Sun and P. Chen (2015). "Graphene quantum dots for ultrasensitive detection of acetylcholinesterase and its inhibitors." *2D Materials* 2(3): 034018.

Li, R., S. Wang, Y. Wang and K. Yu (2018). "Development of a novel methodology for in vivo quantification of N/O/S-containing polycyclic aromatic hydrocarbons located on the epidermis of mangrove roots using graphene quantum dots as a fluorescence quencher." *Marine Pollution Bulletin* 127: 424-428.

Li, R., X. Wang, Z. Li, H. Zhu and J. Liu (2018). "Folic acid-functionalized graphene quantum dots with tunable fluorescence emission for cancer cell imaging and optical detection of Hg^{2+}." *New Journal of Chemistry* 42(6): 4352-4360.

Li, R., L. Liu, H. Zhu and Z. Li (2018). "Synthesis of gold-palladium nanowaxberries/dodecylamine-functionalized graphene quantum dots-graphene micro-aerogel for voltammetric determination of peanut allergen Ara h 1." *Analytica Chimica Acta* 1008: 38-47.

Li, X. Y., Q. Zhang, N. Wang, J.-J. Liu and J. Wang (2020). "Cu^{2+}-mediated Fluorescence Switching of Graphene Quantum Dots for Highly Selective Detection of Glutathione." *Chinese Journal of Analytical Chemistry* 48(3): 339-346.

Li, Y., L. Sun, J. Qian, C. Wang, Q. Liu, E. Han, N. Hao, L. Zhang, J. Cai and K. Wang (2016). "A homogeneous assay for highly sensitive detection of CaMV35S promoter in transgenic soybean by förster resonance energy transfer between nitrogen-doped graphene quantum dots and Ag nanoparticles." *Analytica Chimica Acta* 948: 90-97.

Liang, R. P., W. B. Qiu, H. F. Zhao, C. Y. Xiang and J. D. Qiu (2016). "Electrochemiluminescence resonance energy transfer between graphene quantum dots and graphene oxide for sensitive protein kinase activity and inhibitor sensing." *Analytica Chimica Acta* 904: 58-64.

Lin, L., X. Song, Y. Chen, M. Rong, Y. Wang, L. Zhao, T. Zhao and X. Chen (2015). "Europium-decorated graphene quantum dots as a fluorescent probe for label-free, rapid and sensitive detection of Cu^{2+} and l-cysteine." *Analytica Chimica Acta* 891: 261-268.

Lin, L., and S. Zhang (2012). "Creating high yield water soluble luminescent graphene quantum dots via exfoliating and disintegrating carbon nanotubes and graphite flakes." *Chemical Communications* 48(82): 10177-10179.

Liu, J., Z. T. Chen, D. S. Tang, Y. B. Wang, L. T. Kang and J. N. Yao (2015). "Graphene quantum dots-based fluorescent probe for turn-on sensing of ascorbic acid." *Sensors and Actuators, B: Chemical* 212: 214-219.

Liu, X., W. Gao, X. Zhou and Y. Ma (2014). "Pristine graphene quantum dots for detection of copper ions." *Journal of Materials Research* 29(13): 1401-1407.

Liu, Y., C. Zhao, A. Sabirsh, L. Ye, X. Wu, H. Lu and J. Liu (2021). "A Novel Graphene Quantum Dot-Based mRNA Delivery Platform." *Chemistry Open* 10(7): 666-671.

Liu, Y., N. Cao, W. Gui and Q. Ma (2018). "Nitrogen-doped graphene quantum dots-based fluorescence molecularly imprinted sensor for thiacloprid detection." *Talanta* 183: 339-344.

Lu, F., Y. H. Zhou, L. H. Wu, J. Qian, S. Cao, Y. F. Deng and Y. Chen (2019). "Highly Fluorescent Nitrogen-Doped Graphene Quantum Dots' Synthesis and Their Applications as Fe(III) Ions Sensor." *International Journal of Optics* 2019: 8724320.

Lu, H., W. Li, H. Dong and M. Wei (2019). "Graphene Quantum Dots for Optical Bioimaging." *Small* 15(36): 1902136.

Lu, Q., W. Wei, Z. Zhou, Z. Zhou, Y. Zhang and S. Liu (2014). "Electrochemiluminescence resonance energy transfer between graphene quantum dots and gold nanoparticles for DNA damage detection." *Analyst* 139(10): 2404-2410.

Luo, P., Z. Ji, C. Li and G. Shi (2013). "Aryl-modified graphene quantum dots with enhanced photoluminescence and improved pH tolerance." *Nanoscale* 5: 7361-7367.

Mahmoud, A. M., M. M. El-Wekil, M. H. Mahnashi, M. F. B. Ali and S. A. Alkahtani (2019). "Modification of N,S co-doped graphene quantum dots with p-aminothiophenol-functionalized gold nanoparticles for molecular imprint-based voltammetric determination of the antiviral drug sofosbuvir." *Microchimica Acta* 186(9): 617.

Mahmoudpour, M., A. Saadati, M. Hasanzadeh and H. Kholafazad-kordasht (2021). "A stretchable glove sensor toward rapid monitoring of trifluralin: A new platform for the on-site recognition of herbicides based on wearable flexible sensor technology using lab-on-glove." *Journal of Molecular Recognition* 34(10): 2923.

Malekzad, H., M. Hasanzadeh, N. Shadjou and A. Jouyban (2017). "Highly sensitive immunosensing of prostate specific antigen using poly cysteine caped by graphene quantum dots and gold nanoparticle: A novel signal amplification strategy." *International Journal of Biological Macromolecules* 105: 522-532.

Mandal, D., S. Khatun, A. N. Gupta and A. Chandra (2019). "DNA supported graphene quantum dots for Ag ion sensing." *Nanotechnology* 30(25) 255501..

Mansuriya, B. D., and Z. Altintas (2020). "Graphene Quantum Dot-Based Electrochemical Immunosensors for Biomedical Applications." *Materials* (Basel) 13(1): 96.

Mao, H., Z. Zuo, N. Yang, J. S. Huang and Y. Yan (2017). "A microfluidic colorimetric biosensor for chlorpyrifos determination based on peroxidase-like $CuFe_2O_4$/GQDs magnetic nanoparticles." *Journal of Residuals Science and Technology* 14(1): 255-269.

Mao, Y., C. Zhao, S. Ge, T. Luo, J. Chen, J. Liu, F. Xi and J. Liu (2019). "Gram-scale synthesis of nitrogen doped graphene quantum dots for sensitive detection of mercury ions and L-cysteine." *RSC Advances* 9(57): 32977-32983.

Markovic, Z. M., S. P. Jovanovic, P. Z. Maskovic, M. Danko, M. Micusik, V. B. Pavlovic, D. D. Milivojevic, A. Kleinova, Z. Spitalsky and B. M. T. Markovic (2018). "Photo-induced antibacterial activity of four graphene based nanomaterials on a wide range of bacteria." *RSC Advances* 8(55): 31337-31347.

Markovic, Z., L. M. Harhaji-Trajkovic, B. M. Todorovic-Markovic, D. P. Kepic, K. M. Arsikin, S. P. Jovanovic, A. C. Pantovic, M. D. Dramicanin and V. S. Trajkovic (2011). "In vitro comparison of the photothermal anticancer activity of graphene nanoparticles and carbon nanotubes." *Biomaterials* 32(4): 1121-1129.

Meht, J., N. Bhardwaj, S. K. Bhardwaj, S. K. Tuteja, P. Vinayak, A. K. Paul, K.-H. Kim and A. Deep (2017). "Graphene quantum dot modified screen printed immunosensor for the determination of parathion." *Analytical Biochemistry* 523: 1-9.

Milenković, M., A. Mišović, D. Jovanović, A. Popović Bijelić, G. Ciasca, S. Romanò, A. Bonasera, M. Mojsin, J. Pejić, M. Stevanović and S. Jovanović (2021). "Facile Synthesis of L-Cysteine Functionalized Graphene Quantum Dots as a Bioimaging and Photosensitive Agent." *Nanomaterials* 11(8): 1879.

Monroe, J. D., E. Belekov, A. O. Er and M. E. Smith (2019). "Anticancer Photodynamic Therapy Properties of Sulfur-Doped Graphene Quantum Dot and Methylene Blue Preparations in MCF-7 Breast Cancer Cell Culture." *Photochem Photobiol* 95(6): 1473-1481.

Na, W., Z. Qu, X. Chen and X. Su (2018). "A turn-on fluorescent probe for sensitive detection of sulfide anions and ascorbic acid by using sulfanilic acid and glutathione functionalized graphene quantum dots." *Sensors and Actuators B: Chemical* 256: 48-54.

Nagraik, R., A. Kaushal, S. Gupta, S. Sethi, A. Sharma and D. Kumar (2020). "Nanofabricated versatile electrochemical sensor for Leptospira interrogans detection." *Journal of Bioscience and Bioengineering* 129(4): 441-446.

Nair, R. V., R. T. Thomas, A. P. Mohamed and S. Pillai (2020). "Fluorescent turn-off sensor based on sulphur-doped graphene quantum dots in colloidal and film forms for the ultrasensitive detection of carbamate pesticides." *Microchemical Journal* 157: 104971.

Nair, R., P. R. Chandran, A. P. Mohamed and S. Pillai (2021). "Sulphur-doped graphene quantum dot based fluorescent turn-on aptasensor for selective and ultrasensitive detection of omethoate." *Analytica Chimica Acta* 1181: 338893.

Nie, G., Y. Wang, Y. Tang, D. Zhao and Q. Guo (2018). "A graphene quantum dots based electrochemiluminescence immunosensor for carcinoembryonic antigen detection using poly(5-formylindole)/reduced graphene oxide nanocomposite." *Biosensors and Bioelectronics* 101: 123-128.

Niu, X., Y. Zhong, R. Chen, F. Wang, Y. Liu and D. Luo (2018). "A "turn-on" fluorescence sensor for Pb^{2+} detection based on graphene quantum dots and gold nanoparticles." *Sensors and Actuators, B: Chemical* 255: 1577-1581.

Niu, X., Y. Zhong, R. Chen, F. Wang, Y. Liu and D. Luo (2018). "A "turn-on" fluorescence sensor for Pb2+ detection based on graphene quantum dots and gold nanoparticles." *Sensors and Actuators B: Chemical* 255: 1577-1581.

Novoa-De León, I. C., J. Johny, S. Vázquez-Rodríguez, N. García-Gómez, S. Carranza-Bernal, I. Mendivil, S. Shaji and S. Sepúlveda-Guzmán (2019). "Tuning the luminescence of nitrogen-doped graphene quantum dots synthesized by pulsed laser ablation in liquid and their use as a selective photoluminescence on–off–on probe for ascorbic acid detection." *Carbon* 150: 455-464.

Nsibande, S. A., and P. B. C. Forbes (2020). "Development of a turn-on graphene quantum dot-based fluorescent probe for sensing of pyrene in water." *RSC Advances* 10(21): 12119-12128.

Park, S., K.-S. Lee, G. Bozoklu, W. Cai, S. T. Nguyen and R. S. Ruoff (2008). "Graphene Oxide Papers Modified by Divalent Ions—Enhancing Mechanical Properties via Chemical Cross-Linking." *ACS Nano* 2(3): 572-578.

Pathan, S., M. Jalal, S. Prasad and S. Bose (2019). "Aggregation-induced enhanced photoluminescence in magnetic graphene oxide quantum dots as a fluorescence probe for As(III) sensing." *Journal of Materials Chemistry A* 7(14): 8510-8520.

Peng, D., L. Zhang, R. P. Liang and J. D. Qiu (2018). "Rapid Detection of Mercury Ions Based on Nitrogen-Doped Graphene Quantum Dots Accelerating Formation of Manganese Porphyrin." *ACS Sensors* 3(5): 1040-1047.

Pinto, I. A., C. A. T. Toloza, J. M. S. Almeida, A. R. da Silva, D. G. Larrude and R. Q. Aucélio (2019). "Quantification of neomycin in rubella vaccine by off/on metal ion mediated photoluminescence from functionalized graphene quantum dots." *Spectrochimica Acta - Part A: Molecular and Biomolecular Spectroscopy* 220: 117139.

Ponomarenko, L. A., F. Schedin, M. I. Katsnelson, R. Yang, E. W. Hill, K. S. Novoselov and A. K. Geim (2008). "Chaotic Dirac billiard in graphene quantum dots." *Science* 320(5874): 356-358.

Pothipor, C., J. Jakmunee, S. Bamrungsap and K. Ounnunkad (2021). "An electrochemical biosensor for simultaneous detection of breast cancer clinically related microRNAs based on a gold nanoparticles/graphene quantum dots/graphene oxide film." *Analyst* 146(12): 4000-4009.

Prekodravac, J. R., D. P. Kepić, J. C. Colmenares, D. A. Giannakoudakis and S. P. Jovanović (2021). "A comprehensive review on selected graphene synthesis methods: from electrochemical exfoliation through rapid thermal annealing towards biomass pyrolysis." *Journal of Materials Chemistry* C 9(21): 6722-6748.

Qu, C., D. Zhang, R. Yang, J. Hu and L. Qu (2019). "Nitrogen and sulfur co-doped graphene quantum dots for the highly sensitive and selective detection of mercury ion in living cells." *Spectrochimica Acta - Part A: Molecular and Biomolecular Spectroscopy* 206: 588-596.

Raj, S. K., A. Rajput, H. Gupta, R. Patidar and V. Kulshrestha (2019). "Selective recognition of Fe^{3+} and Cr^{3+} in aqueous medium via fluorescence quenching of graphene quantum dots." *Journal of Dispersion Science and Technology* 40(2): 250-255.

Ristic, B., M. M. Milenkovic, I. R. Dakic, B. M. Todorovic-Markovic, M. S. Milosavljevic, M. D. Budimir, V. G. Paunovic, M. D. Dramicanin, Z. M. Markovic and V. S. Trajkovic (2014). "Photodynamic antibacterial effect of graphene quantum dots." *Biomaterials* 35(15): 4428-4435.

Rodwihok, C., T. V. Tam, W. M. Choi, M. Suwannakaew, S. W. Woo, D. Wongratanaphisan and H. S. Kim (2022). "Preparation and Characterization of Photoluminescent Graphene Quantum Dots from Watermelon Rind Waste for the Detection of Ferric Ions and Cellular Bio-Imaging Applications." *Nanomaterials* 12(4): 702.

Roushani, M., S. Kohzadi, S. Haghjoo and A. Azadbakht (2018). "Dual detection of Malation and Hg (II) by fluorescence switching of graphene quantum dots." *Environmental Nanotechnology, Monitoring and Management* 10: 308-313.

Roushani, M., and F. Shahdost-fard (2019). "Applicability of AuNPs@N-GQDs nanocomposite in the modeling of the amplified electrochemical Ibuprofen aptasensing assay by monitoring of riboflavin." *Bioelectrochemistry* 126: 38-47.

Ruiyi, L., J. Hui, L. Nana, X. Dan and L. Zaijun (2022). "Electrochemical detection of chlorpyrifos in fruits with gold-histidine functionalized graphene quantum dot-graphene hybrid and target-induced DNA cycle amplification." *Sensors and Actuators B: Chemical* 355: 131314.

Sahub, C., T. Tuntulani, T. Nhujak and B. Tomapatanaget (2018). "Effective biosensor based on graphene quantum dots via enzymatic reaction for directly photoluminescence detection of organophosphate pesticide." *Sensors and Actuators, B: Chemical* 258: 88-97.

Sajwan, R. K., and P. R. Solanki (2022). "A hybrid optical strategy based on graphene quantum dots and gold nanoparticles for selective determination of gentamicin in the milk and egg samples." *Food Chemistry* 370: 131312.

Samuei, S., J. Fakkar, Z. Rezvani, A. Shomali and B. Habibi (2017). "Synthesis and characterization of graphene quantum dots/CoNiAl-layered double-hydroxide nanocomposite: Application as a glucose sensor." *Analytical Biochemistry* 521: 31-39.

Santos, C. I., L. Rodríguez-Pérez, G. Gonçalves, C. J. Dias, F. Monteiro, M. d. A. F. Faustino, S. I. Vieira, L. A. Helguero, M. Á. Herranz, N. Martín, M. G. P. M. S. Neves, J. M. G. Martinho and E. M. S. Maçôas (2021). "Enhanced Photodynamic Therapy Effects of Graphene Quantum Dots Conjugated with Aminoporphyrins." *ACS Applied Nano Materials* 4(12): 13079-13089.

Schroeder, K. L., R. V. Goreham and T. Nann (2016). "Graphene Quantum Dots for Theranostics and Bioimaging." *Pharmaceutical Research* 33(10): 2337-2357.

Sebastian, D., R. Kala, K. P. Neethu Parvathy and D. P. Savitha (2021). "A fluorescent probe based on visible light-emitting functionalized Graphene Quantum dots for the sensitive and selective detection of Pb(II) ions." *Journal of Materials Science* 56(32): 18126-18146.

Shadjou, R., M. Hasanzadeh, M. Heidar-poor and N. Shadjou (2018). "Electrochemical monitoring of aflatoxin M1 in milk samples using silver nanoparticles dispersed on α-cyclodextrin-GQDs nanocomposite." *Journal of Molecular Recognition* 31(6) 2699.

Shahat, A., N. Y. Elamin and W. Abd El-Fattah (2022). "Spectrophotometric and Fluorometric Methods for the Determination of Fe(III) Ions in Water and Pharmaceutical Samples." *ACS Omega* 7(1): 1288-1298.

Shan, S. J., Y. Zhao, H. Tang and F. Y. Cui (2017). "A Mini-review of Carbonaceous Nanomaterials for Removal of Contaminants from Wastewater." *IOP Conference Series: Earth and Environmental Science* 68: 012003.

Sharma, A. S., S. Ali, D. Sabarinathan, M. Murugavelu, H. Li and Q. Chen (2021). "Recent progress on graphene quantum dots-based fluorescence sensors for food safety and quality assessment applications." *Comprehensive Reviews in Food Science and Food Safety* 20(6): 5765-5801.

Shen, J., W. Chen, Z. Yang, G. Lv, J. Cao, D. Li and X. Liu (2021). "A Critical Review of Graphene Quantum Dots: Synthesis and Application in Biosensors." *NANO* 16(01): 2130001.

Shinde, D. B., and V. K. Pillai (2013). "Electrochemical Resolution of Multiple Redox Events for Graphene Quantum Dots." *Angewandte Chemie International Edition* 52(9): 2482-2485.

Shtepliuk, I., N. M. Caffrey, T. Iakimov, V. Khranovskyy, I. A. Abrikosov and R. Yakimova (2017). "On the interaction of toxic Heavy Metals (Cd, Hg, Pb) with graphene quantum dots and infinite graphene." *Scientific Reports* 7(1): 3934.

Stefanov, C., C. C. Negut, L. A. D. Gugoasa and J. K. F. van Staden (2020). "Gold nanoparticle-graphene quantum dots nanozyme for the wide range and sensitive electrochemical determination of quercetin in plasma droplets." *Microchimica Acta* 187(11): 611.

Tabaraki, R., and A. Nateghi (2016). "Nitrogen- Doped Graphene Quantum Dots: "turn-off" Fluorescent Probe for Detection of Ag^+ Ions." *Journal of Fluorescence* 26(1): 297-305.

Tabish, T. A., C. J. Scotton, D. C. J Ferguson, L. Lin, A. v. der Veen, S. Lowry, M. Ali, F. Jabeen, M. Ali, P. G. Winyard and S. Zhang (2018). "Biocompatibility and toxicity of graphene quantum dots for potential application in photodynamic therapy." *Nanomedicine* 13(15): 1923-1937.

Tade, R. S., and P. O. Patil (2020). "Green synthesis of fluorescent graphene quantum dots and its application in selective curcumin detection." *Current Applied Physics* 20(11): 1226-1236.

Takemura, K., J. Satoh, J. Boonyakida, S. Park, A. D. Chowdhury and E. Y. Park (2020). "Electrochemical detection of white spot syndrome virus with a silicone rubber disposable electrode composed of graphene quantum dots and gold nanoparticle-embedded polyaniline nanowires." *Journal of Nanobiotechnology* 18(1): 152.

Tam, T. V., and W. M. Choi (2018). "One-pot synthesis of highly fluorescent amino-functionalized graphene quantum dots for effective detection of copper ions." *Current Applied Physics* 18(11): 1255-1260.

Thangadurai, T. D., N. Manjubaashini, D. Nataraj, V. Gomes and Y. I. Lee (2022). "A review on graphene quantum dots, an emerging luminescent carbon nanolights: Healthcare and Environmental applications." *Materials Science and Engineering: B* 278: 115633.

Tian, P., L. Tang, K. S. Teng and S. P. Lau (2018). "Graphene quantum dots from chemistry to applications." *Materials Today Chemistry* 10: 221-258.

Ting, S. L., S. J. Ee, A. Ananthanarayanan, K. C. Leong and P. Chen (2015). "Graphene quantum dots functionalized gold nanoparticles for sensitive electrochemical detection of heavy metal ions." *Electrochimica Acta* 172: 7-11.

Tran, H. L., and R. A. Doong (2019). "Sustainable fabrication of green luminescent sulfur-doped graphene quantum dots for rapid visual detection of hemoglobin." *Analytical Methods* 11(35): 4421-4430.

Tufa, L. T., S. Oh, V. T. Tran, J. Kim, K. J. Jeong, T. J. Park, H. J. Kim and J. Lee (2018). "Electrochemical immunosensor using nanotriplex of graphene quantum dots, Fe_3O_4 and Ag nanoparticles for tuberculosis." *Electrochimica Acta* 290: 369-377.

Valipour, A., and M. Roushani (2017). "Using silver nanoparticle and thiol graphene quantum dots nanocomposite as a substratum to load antibody for detection of hepatitis C virus core antigen: Electrochemical oxidation of riboflavin was used as redox probe." *Biosensors and Bioelectronics* 89: 946-951.

Vermisoglou, E., D. Panáček, K. Jayaramulu, M. Pykal, I. Frébort, M. Kolář, M. Hajdúch, R. Zbořil and M. Otyepka (2020). "Human virus detection with graphene-based materials." Biosensors and Bioelectronics 166: 112436.B. Wang, J. Shen, Y. Huang, Z. Liu and H. Zhuang (2018). "Graphene Quantum Dots and Enzyme-Coupled Biosensor for Highly Sensitive Determination of Hydrogen Peroxide and Glucose." *International Journal of Molecular Sciences* 19(6): 1696.

Wang, C., F. Yang, Y. Tang, W. Yang, H. Zhong, C. Yu, R. Li, H. Zhou, Y. Li and L. Mao (2018). "Graphene quantum dots nanosensor derived from 3D nanomesh graphene frameworks and its application for fluorescent sensing of Cu^{2+} in rat brain." *Sensors and Actuators, B: Chemical* 258: 672-681.

Wang, C., Y. Sun, J. Jin, Z. Xiong, D. Li, J. Yao and Y. Liu (2018). "Highly selective, rapid-functioning and sensitive fluorescent test paper based on graphene quantum dots for on-line detection of metal ions." *Analytical Methods* 10(10): 1163-1171.

Wang, D., P. Wang, D. Liu and Z. Zhou (2019). "Fluorometric atrazine assay based on the use of nitrogen-doped graphene quantum dots and on inhibition of the activity of tyrosinase." *Microchimica Acta* 186(8): 527.

Wang, F., X. Fu, X. Chai, Q. Han, H. Wang and Q. Hao (2021). "Highly selective fluorometric detection of para-nitrophenol from its isomers by nitrogen-doped graphene quantum dots." *Microchemical Journal* 168: 106389.

Wang, F., Z. Gu, W. Lei, W. Wang, X. Xia and Q. Hao (2014). "Graphene quantum dots as a fluorescent sensing platform for highly efficient detection of copper(II) ions." *Sensors and Actuators B: Chemical* 190: 516-522.

Wang, L., J. Zheng, Y. Li, S. Yang, C. Liu, Y. Xiao, J. Li, Z. Cao and R. Yang (2014). "AgNP-DNA@GQDs hybrid: New approach for sensitive detection of H_2O_2 and glucose via simultaneous AgNP etching and DNA cleavage." *Analytical Chemistry* 86(24): 12348-12354.

Wang, Q., L. Li, T. Wu, X. Kong, Q. Ma and C. Ma (2020). "A graphene quantum dots-Pb^{2+} based fluorescent switch for selective and sensitive determination of D-penicillamine." *Spectrochimica Acta - Part A: Molecular and Biomolecular Spectroscopy* 229: 117924.

Wang, Z., D. Chen, B. Gu, B. Gao, Z. Liu, Y. Yang, Q. Guo, X. Zheng and G. Wang (2020). "Yellow emissive nitrogen-doped graphene quantum dots as a label-free fluorescent probe for Fe^{3+} sensing and bioimaging." *Diamond and Related Materials* 104: 107749.

Wolk, A., M. Rosenthal, S. Neuhaus, K. Huber, K. Brassat, J. K. N. Lindner, R. Grothe, G. Grundmeier, W. Bremser and R. Wilhelm (2018). "A Novel Lubricant Based on Covalent Functionalized Graphene Oxide Quantum Dots." *Scientific Reports* 8(1): 5843.

Xiaoyan, Z., L. Ruiyi, L. Zaijun, G. Zhiguo and W. Guangli (2016). "Ultrafast synthesis of gold/proline-functionalized graphene quantum dots and its use for ultrasensitive electrochemical detection of P-acetamidophenol." *RSC Advances* 6(48): 42751-42755.

Xu, D., R. Li, G. Wang, H. Zhu and Z. Li (2021). "Electrochemical detection of carbendazim in strawberry based on a ruthenium-graphene quantum dot hybrid with a three-dimensional network structure and Schottky heterojunction." *New Journal of Chemistry* 45(45): 21308-21314.

Xu, Q., H. Yuan, X. Dong, Y. Zhang, M. Asif, Z. Dong, W. He, J. Ren, Y. Sun and F. Xiao (2018). "Dual nanoenzyme modified microelectrode based on carbon fiber coated with AuPd alloy nanoparticles decorated graphene quantum dots assembly for electrochemical detection in clinic cancer samples." *Biosensors and Bioelectronics* 107: 153-162.

Xu, Y., S. Wang, X. Hou, Z. Sun, Y. Jiang, Z. Dong, Q. Tao, J. Man and Y. Cao (2018). "Coal-derived nitrogen, phosphorus and sulfur co-doped graphene quantum dots: A promising ion fluorescent probe." *Applied Surface Science* 445: 519-526.

Xuan, W., L. Ruiyi, F. Saiying, L. Zaijun, W. Guangli, G. Zhiguo and L. Junkang (2017). "D-penicillamine-functionalized graphene quantum dots for fluorescent detection of Fe^{3+} in iron supplement oral liquids." *Sensors and Actuators, B: Chemical* 243: 211-220.

Yan, Q., Y. Yang, Z. Tan, Q. Liu, H. Liu, P. Wang, L. Chen, D. Zhang, Y. Li and Y. Dong (2018). "A label-free electrochemical immunosensor based on the novel signal amplification system of AuPdCu ternary nanoparticles functionalized polymer nanospheres." *Biosensors and Bioelectronics* 103: 151-157.

Yan, Y., Q. Liu, X. Dong, N. Hao, S. Chen, T. You, H. Mao and K. Wang (2016). "Copper(I) oxide nanospheres decorated with graphene quantum dots display improved electrocatalytic activity for enhanced luminol electrochemiluminescence." *Microchimica Acta* 183(5): 1591-1599.

Yan, Y., Q. Liu, X. Du, J. Qian, H. Mao and K. Wang (2015). "Visible light photoelectrochemical sensor for ultrasensitive determination of dopamine based on synergistic effect of graphene quantum dots and TiO_2 nanoparticles." *Analytica Chimica Acta* 853(1): 258-264.

Yang, L., A. Qin, S. Chen, L. Liao, J. Qin and K. Zhang (2018). "Manganese(ii) enhanced fluorescent nitrogen-doped graphene quantum dots: a facile and efficient synthesis and their applications for bioimaging and detection of Hg^{2+} ions." *RSC Advances* 8(11): 5902-5911.

Yang, Y., Q. Liu, Y. Liu, J. Cui, H. Liu, P. Wang, Y. Li, L. Chen, Z. Zhao and Y. Dong (2017). "A novel label-free electrochemical immunosensor based on functionalized nitrogen-doped graphene quantum dots for carcinoembryonic antigen detection." *Biosensors and Bioelectronics* 90: 31-38.

Yang, Y., G. Fang, X. Wang, F. Zhang, J. Liu, W. Zheng and S. Wang (2017). "Electrochemiluminescent graphene quantum dots enhanced by MoS_2 as sensing platform: a novel molecularly imprinted electrochemiluminescence sensor for 2-methyl-4-chlorophenoxyacetic acid assay." *Electrochimica Acta* 228: 107-113.

Yang, Y., T. Zou, Z. Wang, X. Xing, S. Peng, R. Zhao, X. Zhang and Y. Wang (2019). "The Fluorescent Quenching Mechanism of N and S Co-Doped Graphene Quantum Dots with Fe^{3+} and Hg^{2+} Ions and Their Application as a Novel Fluorescent Sensor." *Nanomaterials* 9(5): 738.

Yang, Y., X. Xiao, X. Xing, Z. Wang, T. Zou, Z. Wang, R. Zhao and Y. Wang (2019). "One-pot synthesis of N-doped graphene quantum dots as highly sensitive fluorescent sensor for detection of mercury ions water solutions." *Materials Research Express* 6(9): 095615.

Yang, Y., X. Xiao, X. Xing, Z. Wang, T. Zou, Z. Wang, R. Zhao and Y. Wang (2019). "Rhodamine B assisted graphene quantum dots flourescent sensor system for sensitive recognition of mercury ions." *Journal of Luminescence* 207: 273-281.

Yanyan, Z., J. Lin, L. Xie, H. Tang, K. Wang and J. Liu (2021). "One-Step Preparation of Nitrogen-Doped Graphene Quantum Dots With Anodic Electrochemiluminescence for Sensitive Detection of Hydrogen Peroxide and Glucose." *Frontiers in Chemistry* 9: 688358.

Yao, J., and L. Wang (2021). "Graphene quantum dots as nanosensor for rapid and label-free dual detection of Cu^{2+} and tiopronin by means of fluorescence "on-off-on" switching: mechanism and molecular logic gate." *New Journal of Chemistry* 45(44): 20649-20659.

Younis, M. R., G. He, J. Lin and P. Huang (2020). "Recent Advances on Graphene Quantum Dots for Bioimaging Applications." *Frontiers in chemistry* 8: 424-424.

Yousefzadeh, A., J. Abolhasani, J. Hassanzadeh and M. H. Somi (2019). "A highly efficient chemiluminescence system based on an enhancing effect of Ag nanoclusters/graphene quantum dots mixture for ultrasensitive detection of rabeprazole." *Analytical Sciences* 35(4): 385-391.

Yousefzadeh, A., J. Abolhasani, J. Hassanzadeh and M. H. Somi (2019). "Ultrasensitive chemiluminescence assay for cimetidine detection based on the synergistic improving effect of Au nanoclusters and graphene quantum dots." *Luminescence* 34(2): 261-271.

Yu, Z., W. Ma, T. Wu, J. Wen, Y. Zhang, L. Wang, Y. He, H. Chu and M. Hu (2020). "Coumarin-Modified Graphene Quantum Dots as a Sensing Platform for Multicomponent Detection and Its Applications in Fruits and Living Cells." *ACS Omega* 5(13): 7369-7378.

Zanut, A., A. Fiorani, S. Canola, T. Saito, N. Ziebart, S. Rapino, S. Rebeccani, A. Barbon, T. Irie, H.-P. Josel, F. Negri, M. Marcaccio, M. Windfuhr, K. Imai, G. Valenti and F. Paolucci (2020). "Insights into the mechanism of coreactant electrochemiluminescence facilitating enhanced bioanalytical performance." *Nature Communications* 11(1): 2668.

Zhang D., Z. Wu and X. Zong (2019). "Metal-organic frameworks-derived zinc oxide nanopolyhedra/S, N: graphene quantum dots/polyaniline ternary nanohybrid for high-performance acetone sensing." *Sensors and Actuators, B: Chemical* 288: 232-242.

Zhang, C., B. Lin, Y. Cao, M. Guo and Y. Yu (2017). "Fluorescence Determination of Omethoate Based on a Dual Strategy for Improving Sensitivity." *Journal of Agricultural and Food Chemistry* 65(14): 3065-3073.

Zhang, Y., J. Qi, M. Li, D. Gao and C. Xing (2021). "Fluorescence Probe Based on Graphene Quantum Dots for Selective, Sensitive and Visualized Detection of Formaldehyde in Food." *Sustainability* 13(9): 5273.

Zhang, Y., C. Yang, D. Yang, Z. Shao, Y. Hu, J. Chen, L. Yuwen, L. Weng, Z. Luo and L. Wang (2018). "Reduction of graphene oxide quantum dots to enhance the yield of reactive oxygen species for photodynamic therapy." *Physical Chemistry Chemical Physics* 20(25): 17262-17267.

Zhang, Y., Q. Rong, J. Zhao, J. Zhang, Z. Zhu and Q. Liu (2018). "Boron-doped graphene quantum dot/Ag-$LaFeO_3$ p-p heterojunctions for sensitive and selective benzene detection." *Journal of Materials Chemistry A* 6(26): 12647-12653.

Zhou, C., W. Jiang and B. K. Via (2014). "Facile synthesis of soluble graphene quantum dots and its improved property in detecting heavy metal ions." *Colloids and Surfaces B: Biointerfaces* 118: 72-76.

Zhou, C., Y. Chen, X. You, Y. Dong and Y. Chi (2017). "An Electrochemiluminescent Biosensor Based on Interactions between a Graphene Quantum Dot−Sulfite Co-reactant System and Hydrogen Peroxide." *Chem Electro Chem* 4(7): 1783-1789.

Zhu, S., Y. Song, J. Wang, H. Wan, Y. Zhang, Y. Ning and B. Yang (2017). "Photoluminescence mechanism in graphene quantum dots: Quantum confinement effect and surface/edge state." *Nano Today* 13: 10-14.

Zhu, W., M. Saddam Khan, W. Cao, X. Sun, H. Ma, Y. Zhang and Q. Wei (2018). "$Ni(OH)_2$/NGQDs-based electrochemiluminescence immunosensor for prostate specific antigen detection by coupling resonance energy transfer with Fe_3O_4@MnO_2 composites." *Biosensors and Bioelectronics* 99: 346-352.

Zhu, X., Y. Liu, P. Li, Z. Nie and J. Li (2016). "Applications of graphene and its derivatives in intracellular biosensing and bioimaging." *Analyst* 141(15): 4541-4553.

Zhu, X., Z. Zhang, Z. Xue, C. Huang, Y. Shan, C. Liu, X. Qin, W. Yang, X. Chen and T. Wang (2017). "Understanding the Selective Detection of Fe^{3+} Based on Graphene

Quantum Dots as Fluorescent Probes: The K_{sp} of a Metal Hydroxide-Assisted Mechanism." *Analytical Chemistry* 89(22): 12054-12058.

Zhu, Y., S. Lu, A. Gowri Manohari, X. Dong, F. Chen, W. Xu, Z. Shi and C. Xu (2017). "Polydopamine interconnected graphene quantum dots and gold nanoparticles for enzymeless H_2O_2 detection." *Journal of Electroanalytical Chemistry* 796: 75-81.

Zor, E., E. Morales-Narváez, A. Zamora-Gálvez, H. Bingol, M. Ersoz and A. Merkoçi (2015). "Graphene quantum dots-based photoluminescent sensor: A multifunctional composite for pesticide detection." *ACS Applied Materials and Interfaces* 7(36): 20272-20279.

Zor, E., E. Morales-Narváez, A. Zamora-Gálvez, H. Bingol, M. Ersoz and A. Merkoçi (2015). "Graphene Quantum Dots-based Photoluminescent Sensor: A Multifunctional Composite for Pesticide Detection." *ACS Applied Materials & Interfaces* 7(36): 20272-20279.

Zou, L., Z. Gu and M. Sun (2015). "Review of the application of quantum dots in the heavy-metal detection." *Toxicological and Environmental Chemistry* 97:1-24.

Chapter 3

Revealing the Magical Role of Graphene Quantum Dots in Biological and Energy Related Applications

Deepa Sebastian
Athira Pallikkara
and Kala Ramakrishnan*
Department of Applied Chemistry, Cochin University of Science and Technology, Kochi, Kerala, India

Abstract

With the developments in nano sciences, quantum dots are known to be at remarkable position of nanotechnology field. Being a novel and promising zero-dimensional material of the carbon nano family, Graphene Quantum Dots (GQDs) have been widely explored owing to their exceptional and beneficial structural, optical, chemical, physical, electrical, and biological features. The distinctive size-dependent properties manifested by them are a reflection of coupled structural attributes which arise from graphene core, embellished with quantum confinement and edge effect. This aids them to exhibit versatile applications in biomedical and energy related domains. This chapter focuses on the various promising properties of GQD such as quantum confinement effect, edge effect, bandgap tunability, unique fluorescence features, biocompatibility, and so forth. Travelling through these properties slowly discloses the various structural parameters to modulate their opto-electronic properties. This aura of GQDs influences their

* Corresponding Author's Email: kala@cusat.ac.in.

In: Graphene Quantum Dots (GQDs)
Editor: Aleksey Kuznetsov
ISBN: 979-8-89113-638-0

biological and energy related applications which led us to summarize several recent advances in the areas of bio-sensing, bio-imaging, solar cells, and light-emitting diodes.

Keywords: graphene quantum dots, optical properties, bioimaging, biosensing, solar cells, light-emitting diodes

1. Introduction

In the beginning of 21st century (in 2004) graphene, an exquisite and novel invention, was made by Andre Geim and Konstantin Novoselov [1]. It is enriched with high thermal/electrical conductivity, elasticity, flexibility, hardness, and higher strength than steel. With its discovery, the various features and the utilization of them in sundry fields have been intensively deliberated and accomplished. Despite all these charms, graphene possesses some inherent properties such as zero-bandgap, low solubility, low absorptivity and so forth, which may not be suitable for certain applications. To conquer these blemishes, structural modifications of graphene were consequently explored [2].

Later graphene quantum dots (GQDs) emerged in 2008, which is the beginning of a new era of biocompatible quantum dots. The discovery of GQDs harmonizes with their prior member, namely, carbon dots, but structurally both are different. GQDs consist of sp^2 hybridized graphene lattice inside a dot with edge terminated by various oxygenated functionalities such as hydroxyl (-OH), carboxyl (-COOH), etc., which assists the water solubility and further functionalization. GQDs are generally less than 10 layers thick with size between 1-100 nm. Moreover, GQDs comprise of several unique and novel attributes. They exhibit good crystallinity and similar lattice structure as graphene, but their intrinsic properties get different from carbonaceous 0D materials like carbon dots and carbon nanoparticles [3].

Several methods are available for the synthesis of GQDs, they are the top-down and bottom-up methods. Avenues of top-down methods originated from the oxidative cleavage of carbonaceous precursors, progresses through simple operations, and ends up with large-scale production. Methods involved are acidic exfoliation, hydrothermal synthesis, and electrochemical oxidation and so on. Bottom-up routes produce well-defined GQDs with respect to shape and size, even so necessitate complex synthesis methods that are strenuous to use, lead to low yields and high cost of production. Both methods have their own advantages and disadvantages. The mass production of GQDs have many limitations, but this is a significant factor required for GQDs in their industrial applications [4-6].

Moving to their properties, two significant effects, namely, quantum confinement and edge effect, play an important role in determining the opto-electronic properties of GQDs. Graphene depicts infinite exciton Bohr radius. When this material is converted to zero-dimensional one, the redistribution of electrons opens up bandgap, and subsequent quantum confinement effect take place in GQDs. Unlike conventional semiconductor QDs, GQDs possess unique photoluminescence (PL) properties, in which up-conversion luminescence and excitation dependent PL are categorized as unique features like two sides of a coin [7]. PL mechanism of GQDs can be explained to an extent with the aid of their size, functional groups, electronic energy levels, etc.; there is no universal agreement on specific mechanism. Influenced by the quantum confinement effect (QCE) and edge effect, GQDs exhibit non-toxicity, solubility, biocompatibility, stable PL, convenient quantum yield, surface grafting, inertness, and so forth [8]. Recently, numerous charming optoelectronic properties of GQDs were revealed and their potential applications in biological, energy harvesting, and storage devices have been explored [9]. But the GQDs research is still at its early stage and numerous disputes must be addressed. In comparison to conventional semiconductor quantum dots, GQDs provide low quantum yield values. Recent studies showed that the low quantum yield value of GQDs can be overcome by surface modification such as functionalization, nanocomposites preparation, and so forth [10].

Along with this, the major fluorescent properties such as optical bandgap and photoluminescence stability are mainly dependent on the size of the GQDs. Thus, alteration in the intrinsic attributes of GQDs can be changed by controlling its size. The experimental evidence reported for the contemporary mechanism for the PL properties of GQDs lacks some logical validity. Hence, experimental studies coupled with theoretical studies are used to get an insight into the PL mechanism of GQDs [11]. The narrow spectral coverage of GQDs is yet another problem [12]. Nowadays, extending the PL spectra of GQDs to visible and near IR region is the most challenging task. The forthcoming modifications on GQD will be auspicious ones when the above disputes are clarified. Even though GQDs face some limitations, their boundless features would benefit numerous applications in diverse fields. Tailoring their unique PL properties with the aid of several strategies such as functionalization, doping, hybrid and nanocomposites formation and so forth is now under scrutiny [13]. In the succeeding years GQDs may be admired as the highly demanding candidate at the cutting-edge of modern technology. In this chapter we are dealing with fascinating properties of GQDs and their biological as well as energy-related applications.

2. Structure of GQDs

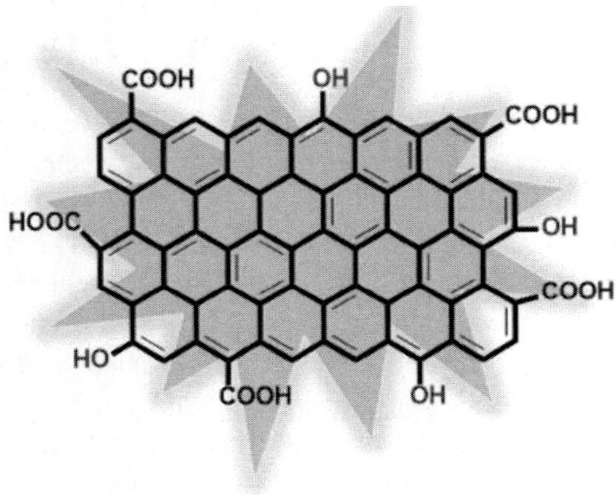

Figure 1. Structure of Graphene Quantum Dots.

Structurally GQDs simulate the crystalline structure of single or multiple layers of graphene. Typically, the most reported GQDs are circular or elliptical in shape, however, triangular, hexagonal, and quadrate GQDs are also reported [14]. The core of GQD is consisted of sp^2 hybridized aromatic rings with edge enriched with various oxygenated functionalities such as hydroxyl (-OH), carboxyl (-COOH) groups etc. Unlike carbon quantum dots (CQDs), GQDs can have lateral dimension greater than 10 nm [15]. The largest reported lateral dimension is approximately 60 nm. But analogous to CQDs, GQDs also depict lattice spacing in the range of 0.18-0.24 nm [16]. Due to the occurrence of abundant functionalities at the edge, the graphitic interlayer spacing of GQD is found to be 0.34 nm [17]. Various techniques such as high-resolution transmission electron microscopy (HR-TEM), X-ray diffraction pattern, (XRD) X-ray photoelectron spectroscopy (XPS), and infrared spectroscopy (IR) are used for the structural confirmation and functional group determination of GQD [18]. Based on the synthetic strategies, numerous heteroatoms, defects, and functional groups are established on GQDs, which may predominantly revamp the structural and physicochemical properties of these nanodots. Based on this idea an approximate structure of GQD is given in Figure 1.

3. Properties of Graphene Quantum Dots

3.1. Quantum Confinement Effect (QCE)

The combination of 'Quantum' and 'Confinement' gave birth to the idiom 'Quantum Confinement' which means 'a discrete quantity of energy (electrons/excitons) is spatially confined' [19]. QCE is manifested by those particles whose sizes are possible to compare with the wavelength of electron and provides an identity to each nanoparticle. As per the trend followed by QCE, when the size of a particle decreases to nanoscale, its energy levels become discrete, and the bandgap increases. That is, the size of a nanoparticle is inversely proportional to its corresponding bandgap (detailed representation is given in Figure 2). GQDs belong to this class of nanoparticles, which satisfies the above requirements and materializes the phenomenon of QCE. Graphene, which is the precursor of GQDs, possesses zero bandgap, which becomes non-zero by the formation of graphene quantum dots and depends on their method of formation [20, 21].

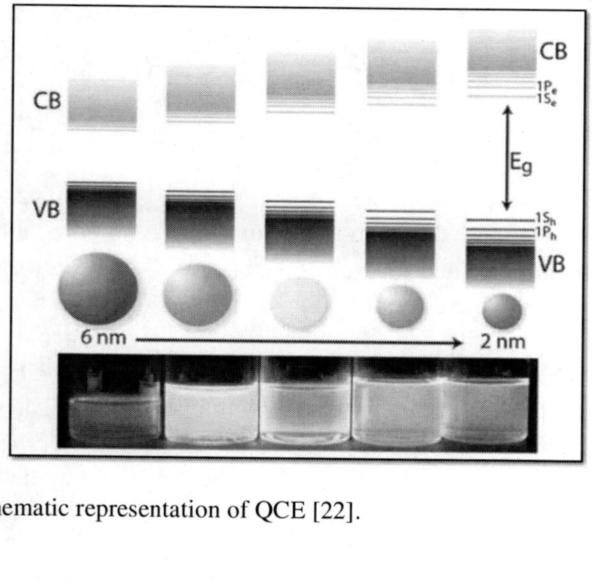

Figure 2. Schematic representation of QCE [22].

3.2. Edge Effect

As the name suggests, edge effect means the modulations in the optical and electronic properties of a nanomaterial due to the peculiarities of its edge. In a general sense this effect is of two types, one is due to the possible shapes of edge and the other is due to the presence of various functional groups, which are discussed as follows. First effect arises from the characteristic shapes of the edges of GQDs. Generally, they are of two type, armchair and zigzag, that is carbyne- and carbene-like edges, respectively, as shown in Figure 3 [23]. If the quantum dots are not well-developed, other shapes are also observed.

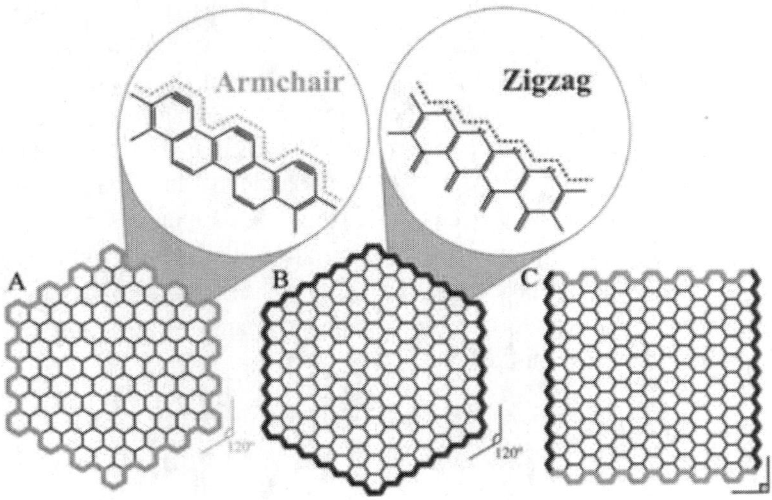

Figure 3. Possible edge geometries of Graphene Quantum Dots [23].

The second one is due to the presence of edge groups. That is, the classic model of GQDs (Figure 1) embodied with a graphene core and a number of edge functionalities. Since the surface states of GQDs are originated from these edge groups which control a major part of the photoluminescence of GQDs, the study of the edge functionalities is an inevitable part of GQD research [24]. Earlier reports state that the carboxyl (-COOH) and amide (-CONH) edge groups exhibit green photoluminescence (PL), whereas blue emission is from hydroxyl groups [23, 24]. According to Zhu et al. all the excited electrons are relaxed to the surface states and show the PL spectrum, irrespective of their LUMO levels [25].

3.3. Optical Properties

3.3.1. Effect of Intrinsic Properties on Optical Features of GQDs

The optical properties of GQDs are mainly derived from their intrinsic features such as size, layers, hybridization, or degree of edge functionalities, which are directly connected to their bandgap tunability. GQDs are comprised of the extended π-π conjugated system that ultimately leads to large surface area and significant surface grafting properties. Structurally GQDs consist of sp^2 hybridized carbon core along with various hydroxyl and carboxyl functionalities at the edge. These functionalities at the edge favour subsequent functionalization with other organic, inorganic, and polymeric systems leading to the tuning of optical properties. Further functionalization can lead to the bandgap tuning and thereby modifications of the PL behaviour. Figures 4 & 5 represent the UV-visible absorption spectra and PL spectra of GQDs and dithienopyrrole derivatives functionalized GQDs [26, 27]. The red shifts observed in the UV-Visible as well as PL spectra indicate that the functionalization is an efficient strategy to improve the photophysical properties of GQDs. Also, the multi-layered GQDs depict UV-visible absorption in a broader range than that of the single-layered GQDs. Along with layer-dependent size, another significant optical property exhibited by GQDs arises from their extended conjugated structure with delocalized π-electrons that leads to a broad-band emission, which is clarified by the following example. The N-GQDs are synthesized via microwave-assisted hydrothermal method using glucose and aqueous ammonia as raw materials, which prove that the layered structure with various degrees of conjugation resulted in versatile PL emission spectra. The PL emission in the UV region and the visible region arises from the localized π-electrons as well as the conjugated π-electrons, respectively [28].

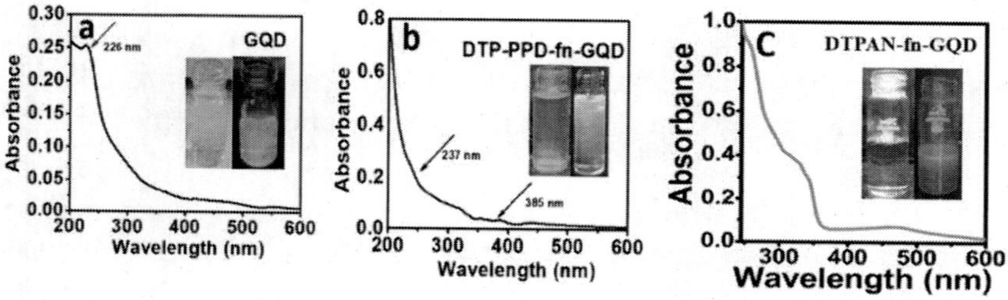

Figure 4. UV-visible absorption spectra of (a) GQD, (b) DTP-PPD-fn-GQD, (c) DTPAN-fn-GQD.

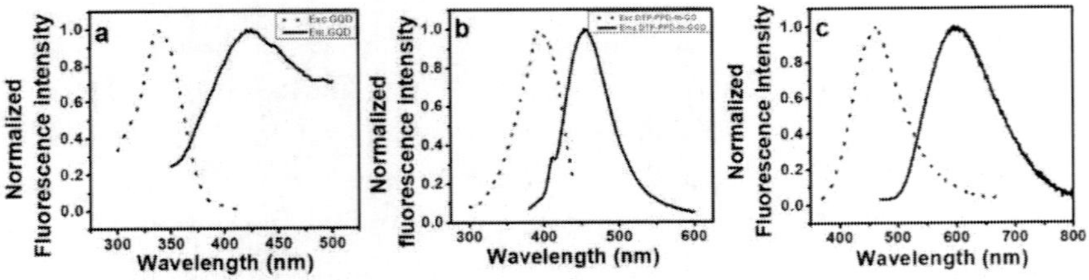

Figure 5. Fluorescence spectra: (a) GQD, (b) DTP-PPD-fn-GQD, (c) DTPAN-fn-GQD.

Unlike conventional semiconductor quantum dots, in GQDs along with the quantum confinement effect the edge effect also plays a major role in determining the optoelectronic properties. As mentioned above, the GQDs are encompassed by the armchair and zigzag edges. Generally, the decrease in the bandgap is related to the localized zigzag edge sites, yet it is broadened by the armchair edges with a blue shift in emission [29]. Thus, the PL properties of GQDs can be tuned by modulating the magnitude of edge sites. The modulation of the edge sites can be achieved via various strategies, among which functionalization is known to be an important one. Functionalizing the edge sites of GQD using small molecules to macromolecules will significantly affect the magnitude of the electron density at the edge site. One such system based on activated GQDs with an edge rich in oxygenated and carboxyl functionalities is illustrated below. Here the activation of GQDs (aGQDs) is done via the ultrasonic method. The edge structure of these GQDs comprises both free and bound edges that act as efficient active sites, which results in superior and tunable fluorescence properties in contrast with non-activated analogues. This approach provides an insight into the modification of GQDs enabling its application towards the light-emitting diodes, optoelectronics, bio-imaging, and so forth [30]. Besides that the shape of GQDs has an influence on the optical properties, typically the shapes of GQDs are influenced by the size and edge sites. The synergy between the size, edge sites, and shape significantly contributes to the tunable optical properties of GQDs.

3.3.2. Surface States Modulation

An important addition to the surface states of GQDs is the surface oxidation states which are induced by the edge functional groups [31]. These surface oxidation states create a series of emissive trap states. The usually greater magnitude of surface trapping leads to a red shift in the PL spectra of GQDs. For a time, it was thought that GQDs can emit only excitation-independent PL spectra as in the conventional fluorophore or small organic molecules [32]. But the discovery of trap states in GQDs led to the exploration of surface state-mediated PL centres that results in an unusual phenomenon called excitation-dependent PL spectra [33]. Moreover, the surface states can significantly influence the charge transfer interaction between the various incoming species. Figure 6 demonstrate the idea behind the evolution of surface trap states in pristine GQDs, and the effect of electron-donating and electron-withdrawing moieties on the exciton dynamics can be observed. In spite of the fact that the presence of oxygen-containing groups contributes to the green or red fluorescence of GQDs, these are susceptible to bringing about the non-radiative recombination of localized electron-hole (e–h) pair, which will hinder the intrinsic state emission. The reduction of trap states in GQDs can be achieved by surface functionalization which is widely utilized [34-36].

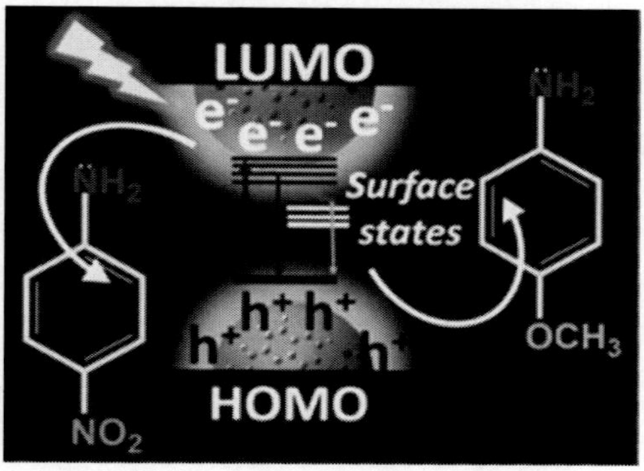

Figure 6. Origin of surface trap states in GQDs and the effect of electron-withdrawing and electron-donating groups on the excited charge transfer interaction between GQDs. Reproduced with permission from ref. [37].

Several spectroscopic studies have been carried out to monitor the surface trap states in pristine GQD, among which transient absorption (TA) spectroscopy earned considerable attention due to its ability to monitor the dynamics of photoexcited excitons in ultrafast time scale [37]. For example, Figure 7 depicts the TA spectra of GQD under photoexcitation at 320 nm, which gives a positive signal indicating that the major portion of GQD consists of surface trap states. The surface functionalization approach provides a platform to tune the structure and optical properties of GQDs. Amongst the various compounds for functionalization the diamines are the preferred choice. It is due to the unique protonation of diamines that significantly enhances the PL properties of a system. An example of the surface passivation done via amine functionalization is given below, which depicts the tunable optical properties on GQDs. The functionalized GQD exhibits highly bright fluorescence and superior quantum yield (about 40%) [36, 37].

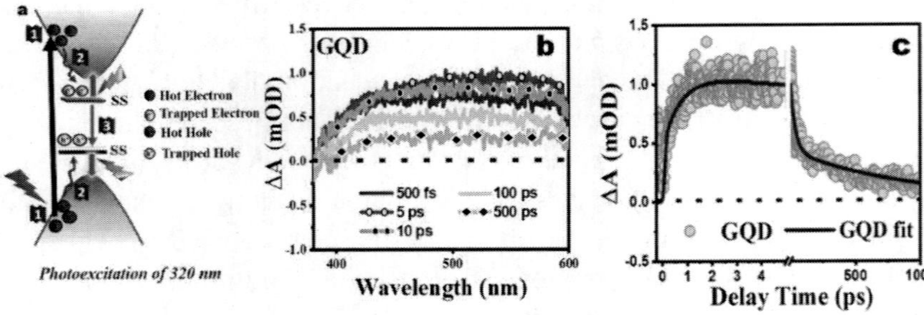

Figure 7. (a) Schematic representation of the photophysical processes occurring in the pristine GQD system under photoexcitation at 320 nm; (1) hot carrier generation, (2) carrier trapping in the surface states (SS), (3) electron-hole recombination. (b) Transient absorption (TA) spectra of GQD spanning the wavelength range of 400 nm - 600 nm at different pump-probe delay times. (c) TA kinetics of GQD monitoring at wavelength 518 nm after photoexcitation of 320 nm (3.88 eV) pump. Reproduced with permission from ref. [37].

3.3.3. Effect of Surface Oxidation/Reduction on Optical Properties

Besides surface functionalization, surface oxidation and reduction approaches are utilized to tune the optical properties of GQDs. Typically, the GQDs obtained via oxidation approaches have high oxygen content. This high degree of oxidation usually lets the system to be red shifted but results in a low quantum yield [38]. A promising alternative to achieve high quantum yield is by lowering the degree of oxidation which in turn results in blue-shifted emission of GQDs. Some reasons for these effects are described as follows, which is defined as the surface state principle. The surface state principle is stated as follows that is, through high degree of surface oxidation the magnitude of surface trap states on GQD is significantly enhanced which in turn results in diminishment of the photoluminescence properties. That is the PL spectra of GQDs originated due to the coupled effect of intrinsic emission and defect state emission, which is attributed to the quantum size effect and edge effect, respectively [39]. Generally, the oxidation of GQDs will create a large number of oxygen groups which subsequently cause a non-radiative transition that results a decrease in the quantum yield values, whereas the surface reduction will promote the transformation of carbonyl, amide and epoxy groups to hydroxyl groups leading to the annihilation of non-radiative recombination and thus the stability of π-conjugated GQDs increases the embellishment of electron density giving rise to high quantum yield [40, 41].

3.4. Biocompatibility of GQDs

Low cytotoxicity and appreciable biocompatibility are the key requirements of a nanomaterial for its bioimaging application. A good candidate for biological application must be nontoxic and adaptable with the human body. Hence the cytotoxicity of GQDs and related materials should be examined before using them in biological applications. There are several factors which influence the biocompatibility of a material [42]. In the case of GQDs, the various carboxylic acid moieties at the edge of GQDs expedite the water solubility and necessary functionalization. This will render GQDs to be low-toxic and eco-friendly substitutes for conventional fluorescent probes.

The biocompatibility of any material can be easily verified through *in vitro* studies. The *in vitro* studies involve testing of biocompatibility of molecular systems on cell lines via various assay methods [43]. This method is comparatively simpler than *in vivo* studies. The following example explains the *in vitro* studies of GQDs on prostate cancer cell lines (DU-145 and PC-3). Here the highest tolerable toxicity for DU-145 cells was obtained for 100 μg/mL of GQDs. Above this value the cell viability decline, meanwhile below this level there is no indication of cell viability. Also, at a concentration of 400 μg/mL, the cell viability retained to 80%, similar observation was seen in PC-3 cell line, leading to the conclusion that GQDs possess good biocompatibility. Likewise, the illustration shown below gives an insight into the membrane permeability of different size GQDs. Here GQDs having two different nano sizes (that is, one small size, ~3 nm, and one large size, ~12 nm) have been synthesized and utilized for cell viability studies. The smaller size GQD entered Madin-Darby canine kidney (MDCK) cells via lipid raft mediated transcytosis mechanism, while the penetration of large size GQD into human stem cells efficiently occurred without any influence on cell viability and proliferation [44, 45].

Based on these properties, we are dealing with the biological and energy related applications in the upcoming sections of this chapter (Figure 8).

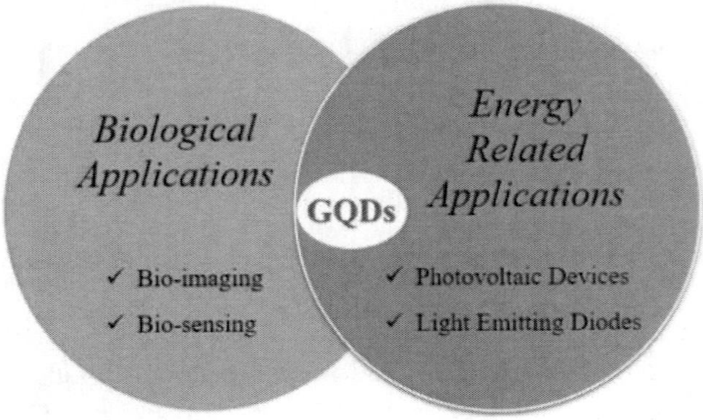

Figure 8. Classification of applications of GQDs under our study.

4. Bioimaging Application of GQDs

4.1. What Is Bioimaging?

Bioimaging is a non-invasive procedure of visualizing the biological activity of biological specimens such as cells, tissue, or gel for a fixed time [46]. It helps to capture the three-dimensional (3D) structures of the specimen without hindering the various physiological life processes. Furthermore, it is frequently used in the medical field to acquire particulars on the 3D structure of the diseased specimen from outside. Bioimaging technique comprises the investigation to glean data concerning the condition of cells, entire biological tissues of subcellular structures, quantification of ions, molecular metabolites, the interaction of live cells, and cells over tissues up to entire multicellular organisms through fluorescence, phosphorescence, X-ray, electrons, positrons, magnetic resonance, and ultrasound as the source for the imaging process. The prerequisite for majority of the bioimaging application is the necessary tracers such as fluorophores and advanced microscopic instruments such as confocal microscope and so forth [47].

4.2. Significance of Optical Bioimaging

The optical bioimaging technique is one of the highly advanced methods that makes use of optical contrast which is induced by the difference in the transmission, reflection, and fluorescence of light in the imaging region and background region [48]. Optical imaging overcomes the majority of the drawbacks of conventional imaging techniques such as X-ray imaging, Magnetic resonance imaging (MRI) scanning, radio isotopic imaging, and ultrasound imaging. This technique utilizes the difference in the optical properties such as reflection, scattering, fluorescence, quantum yield, excitation dependent/independent PL properties, and near-IR excitation-emission of a biological specimen, a cell, tissue, an organ and so forth. The main benefits of the use of various kinds of radiation such as near IR, X-ray, fluorescence and so forth can improve the cell penetration capacity of the probe with a less toxic effect on the cells. The selection of convenient probes plays a major role in bioimaging application since the resolution, sensitivity and versatility of optical microscopy are mainly dependent on the optical properties of the various fluorescent probes [49].

The major advantages of using optical imaging are listed as follows.

- Less harmful
- Imaging of size between the range of 10 nm to macro size is possible.
- Multicolor imaging is obtained with the aid of reflection, transmission, fluorescence, and phosphorescence along with spectroscopic information.
- *In vivo*, *in vitro*, and *ex vivo* imaging is possible.
- A spectrally resolved and dynamic image of cellular process and tissue engineering is possible.
- Feasibility to combine optical imaging with other techniques such as ultrasound techniques.
- There are abundant modalities available in the optical imaging process.

4.3. Fluorescence Bioimaging - A New Era

Among the optical bioimaging techniques, fluorescence imaging is one of the most recent promising approaches [50]. Fluorescence bioimaging is defined as the foremost resource in diagnostic medicine which is also known as one of the highly sensitive, selective, and specific techniques for bioimaging. This mainly involves the effective use of the emission properties of a fluorophore at a specific wavelength. The difference between the excitation energy and emission energy should be significant so that the efficient imaging of the target cells is possible without the interference of other foreign species. The major advantage of using fluorescence microscopy is the effective utilization of the autofluorescence property of a fluorophore. That means, we can label the cells, tissue, or gel with an appropriate molecule or a fluorophore. After the effective illumination of light one can easily identify the spatial distribution over cellular structures. Meanwhile, the enhanced signal-to-noise ratio provides us a way to differentiate the spatial arrangement even at ultra-small concentrations. The most important classifications of fluorescence microscopy are given as follows [51].

- Spatial filtering: confocal microscopy
- Spatially-resolved localized spectroscopy
- Polarization-resolved spectroscopy
- Time-resolved fluorescence lifetime imaging (FLIM)
- Fluorescence resonance energy transfer (FRET) imaging.

More recently various novel bioimaging techniques/mechanisms such as two-photon fluorescence excitation microscopy, super-resolution, fluorescence recovery/redistribution after photobleaching (FRAP), fluorescence resonance energy transfer (FRET) imaging, photoinduced electron transfer (PET) imaging, aggregation-induced emission enhancement (AIEE) imaging and fluorescence lifetime imaging microscopy (FLIM) have been well studied [52, 53].

4.4. GQDs Are a Leading Candidate in Cellular Imaging

The early-stage recognition of diseases helps to enhance the survival rate of patients, which has inspired researchers to fabricate most sensitive, less toxic, fluorescent materials with great specificity. Formerly, organic dyes and inorganic semiconductor quantum dots based fluorophores have been conventionally utilized for cellular visualization and bioimaging. Nevertheless, the photobleaching, poor water solubility, toxicity, and low quantum yield associated with them are the major hurdles to their practical utilization in bioimaging applications. In this respect, GQDs are one of the novel candidates from the carbon nano family embracing an appreciable potential to strongly substitute these fluorophores by virtue of their promising and tunable PL properties, photostability, and significant biocompatibility, thus contributing magical opportunities for bioimaging. The initial manifestation of GQDs as an efficient fluorescent probe was well studied for cellular dynamics and *in vivo* as well as *in vitro* tumor imaging [54].

Pristine GQDs, functionalized GQDs, and their subsequent derivatives displayed numerous benefits in contrast to other inorganic semiconductor quantum dots and fluorophores. Conventional GQDs have manifested superior photoluminescence responses for bioimaging applications [55]. The fascinating optical properties of GQDs are driven by their structural peculiarities which ultimately determine the path toward bioimaging application, as shown earlier by the structure of GQDs, which carry graphene sheets inside the dots, along with different anchoring groups such as -OH and -COOH at the edge offering excellent water solubility and low cytotoxicity compared with conventional QDs. Pristine GQDs exhibit size-dependent PL properties and bandgap tuning features, hence excitation wavelength-dependent PL properties associated with GQD make them the most curious and mysterious candidate from the carbon nano family. There have been numerous publications on applying the optical properties of GQD in bioimaging applications [56]. The upcoming part briefly explains the influence of optical properties of GQD on bioimaging application.

4.5. Up-Conversion Luminescent (UCL) GQDs for Bioimaging

The need for biocompatible up-conversion luminescent nanoparticles (UCLNPs) is worthy of attention since the traditional down-conversion fluorescent nanoparticles (DCLNPs) deteriorate by insignificant signal-to-noise ratio, autofluorescence, and sudden photo damage to living cells [57]. In this respect GQDs proved to be a finer alternative for conventional DCLNPs because of their biological relevancy, higher order up-conversion PL efficiency, and fulfillment of the sensitivity requirements. For any system the up-conversion event is characterized by exciting the lower electronic levels of photons with near IR radiation (NIR light) to produce visible or UV-light radiation at higher electronic levels. Contrary to nonlinear multiphoton absorption happening in conventional QDs and organic fluorophores, UCL in GQDs involves sequential absorption of two or more photons through surface states energy levels by means of diverse mechanisms such as multiphoton absorption, Auger recombination, and phonon absorption, which provides extra energy for the up-conversion emission (generally called anti-Stokes mechanism). It is argued that multiphoton process alone cannot explain the UCL in GQDs owing to the abiding excitation and emission transfer between the surface trap states [58]. Instrumental factors can also have influence on the final responses. A typical up-conversion luminescent experimental setup is given in Figure 9. This instrument was built up on a reversed microscope with a confocal scanning unit. A continuous laser source of ca. 980 nm is directed and concentrated onto the sample specimen. Various studies demonstrated that unmodified GQDs show comparatively better performance at this excitation wavelength, however the required laser excitation wavelength for other GQDs obtained via various structural modifications (doping, functionalization, nanohybrid formation and so forth) deviates from this standard value [59]. Moving to another component called beam expander will then transform this radiation into beams of larger diameters, which are then homogeneously illuminated all over the specimen. The short-pass filter combined with the lens is calibrated in consonance to the different excitation and emission wavelengths of the UCLNPs. Finally, the images will be recorded by the detector coupled with microscope.

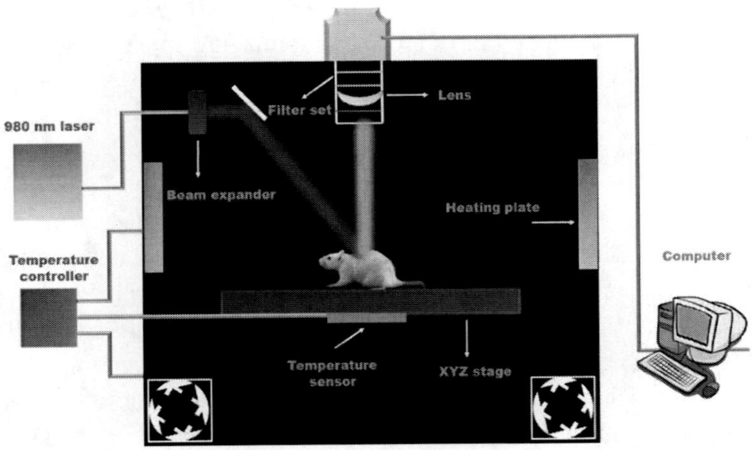

Figure 9. Instrumental setup of the up-conversion fluorescence in vivo imaging system [59].

The UCL based bioimaging technique with GQDs is clarified by the following example [60]. Three categories of GQDs, namely, unmodified GQDs (GQDs), surface modified GQDs (m-GQDs), and reduced GQDs (r-GQDs), are synthesized through various synthetic routes. Here the m-GQDs are obtained through the surface modification via methylamine coupling on GQDs, whereas the r-GQDs are obtained through sodium borohydride ($NaBH_4$) reduction, which allows the specific reduction of the epoxide and carboxylates to hydroxyl groups. These three structurally distinct GQDs show variation in their cytotoxicity and bioimaging performance. Among these the unmodified GQDs and r-GQDs display low *in vitro* cytotoxicity in the methylthiazolyldiphenyltetrazolium bromide (MTT) assay method. Meanwhile the m-GQDs manifested relatively high value of cytotoxicity, that is, about 40% cells remained alive while adding 400 μg of GQDs [60]. It is stated that this discrepancy arose due to the structural modifications carried out on GQDs. The amino groups and methyl chains present on m-GQDs could influence the environment of cells, whereas the abundant hydroxyl functionalities on r-GQDs could serve as a better biocompatible platform that subsequently enhances the survival rate of cells. In addition, all the three GQDs were utilized for the successful UCL imaging of osteoplastic cell lines of mouse (MC3T3-E1). Even though these three kinds of GQDs have their own structural distinctions, all of them undergo rapid penetration and translocation through the cell lines and show highly sensitive and selective bioimaging property. In this case GQD does not undergo any further functionalization or conjugation with fluorophores or other molecules since the unmodified GQD itself is capable of showing highly sensitive and selective bioimaging property. The connected alkyl amines of m-GQDs as well the –OH functionalities of r-GQDs can suppress non-radiative transition. This ultimately enhances the intrinsic state emission, or suppression of defect state emission, resulting in the UCL behaviour. That is, structural modification has a huge impact on the PL behaviour of GQDs and expands their application in bioimaging as well as cytotoxicity performance [61].

4.6. Near-IR Emitting (NIR) GQDs for Bioimaging

The up-conversion PL properties of GQDs are the key feature that are utilized to fabricate near-IR emitting probes based on GQD based materials. Near-IR (NIR) radiation belongs to the electromagnetic radiation family with a longer wavelength than visible light (400-700 nm) (Figure 10). Near-IR emission has been extensively studied to develop bioimaging probes because of its unique advantage of deep penetration into tissues due to their reduced scattering and autofluorescence properties causing less cell damage [62].

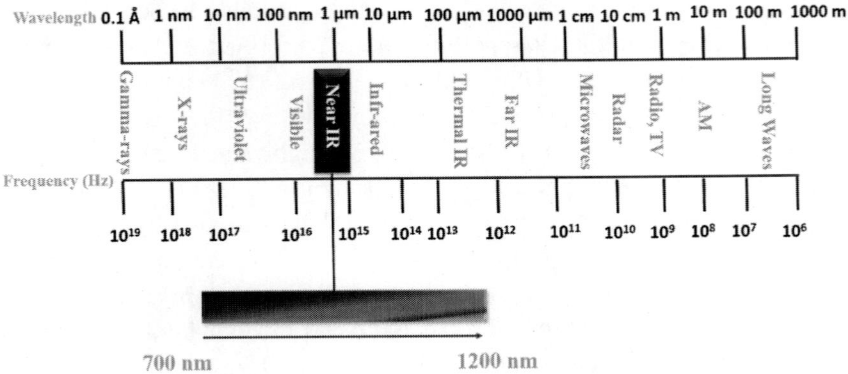

Figure 10. Electromagnetic spectrum and application of NIR light [63].

Unlike fluorescence, NIR emission is associated with multiple exciton dynamics, the majority of which is influenced by triplet state energy. The utilization of triplet energy in typical NIR material is summarized in Figure 11 [63].

- *For a doublet state:* The unpaired electron that lies in the highest single occupied molecular orbital (SOMO) is excited to the lowest single unoccupied molecular orbital (SUMO), and the back electron transfer to SOMO is spin-allowed transition.
- *For a hybridized local and charge transfer state (HLCT):* Here the reverse inter-system crossing (RISC) occurs efficiently which creates a radiative exciton with a ratio greater than 25% through the hot exciton process.
- *For delocalized electronic state:* At this state the ratio between quantum yield of electroluminescence (EL) and photoluminescence (PL) becomes unity when the electron and the hole energies are adequately low.
- *For triplet-triplet annihilation (TTA):* The annihilation between two triplet excitons that generates 62.5% energy for utilization will lead to creation of a higher energy triplet exciton.
- *Thermally activated delayed fluorescence (TADF):* This occurs when the triplet excitons are transferred from the triplet state (T_1) through RISC.
- *Phosphorescence:* This arises from the triplet state to the singlet state (S_0). Theoretical internal quantum efficiency of a phosphorescence process can reach about 100%.

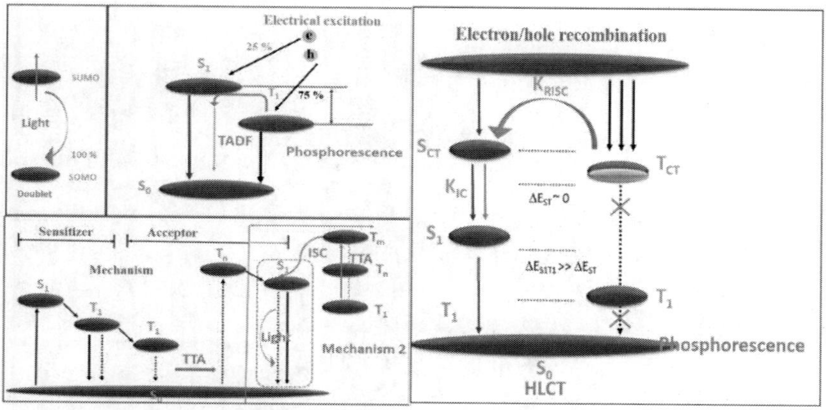

Figure 11. Radiation mechanisms for doublet, fluorescence, thermally activated delayed fluorescence (TADF), phosphorescence, triplet–triplet annihilation (TTA), and hybridized local and charge-transfer (HLCT).

There have been several approaches reported for the fabrication of GQD-based NIR emitting materials [64] Among the various strategies proposed, the reduction in the bandgap (Eg) is one of the most efficient and convenient ways to initiate the NIR emission in GQDs, which can be done by the substituent effect, change of π-conjugation length, hetero atom doping, and developing donor-acceptor (D-A) systems. The above process can significantly influence the HOMO and LUMO energy levels which is the reason behind the rational mechanistic aspects of the bandgap narrowing and NIR emission. Figure 12 shows the two photon (TP) and depth fluorescence images of GQDs modified with CoOOH nanoflakes (CoOOH-NG) applied for HeLa cells for the detection of ascorbic acid (AA). Here both TP and NIR properties are collectively utilized for the endogenous detection of AA in living cells. Here it is seen that by exciting at 810 nm femtosecond laser excitation fluorescence signal is observed at 600 nm and subsequently the tissue images were collected by TP excitation action cross-section ($\delta\Phi$) of 25.12 Goeppert-Mayer (GM). Due to the hybrid type of structure of GQDs designed here an electron transfer (ET) from GQD to CoOOH is feasible suggesting the efficient TP- and NIR-based bioimaging process. Due to the lower fluorescence background exhibited by these GQDs, high-resolution imaging is possible, and hence these systems act as fluorescence labelling agents. Here the effective utilization of two-photon absorption and near-IR fluorescence features with lower fluorescence background is possible [65].

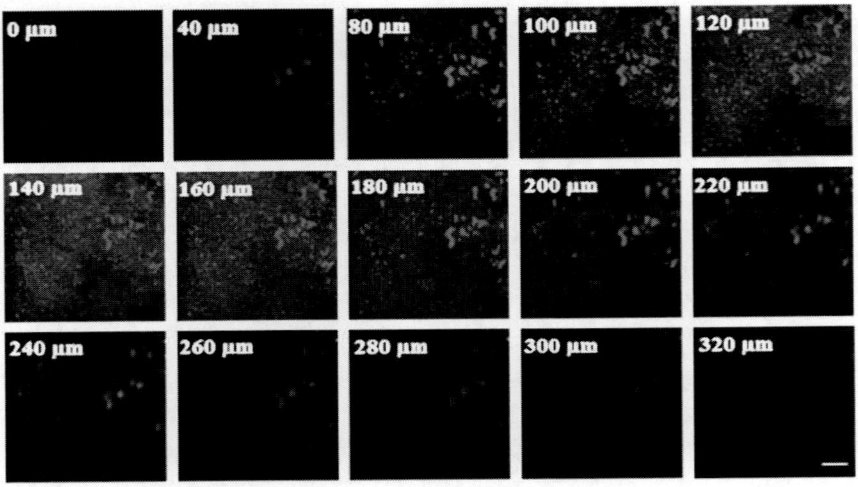

Figure 12. Depth fluorescence images of CoOOH-NGs (200 µg/mL) (the changes of fluorescence intensity with scan depth were determined by spectral confocal multiphoton microscopy in the z-scan mode (step size, 5 µm)). TP images were collected at 580−680 nm and 400−500 nm, respectively. Scale bar = 100 µm. Reproduced with permission from ref. [65].

In another case influence of preparation strategy on the NIR feature and subsequent bioimaging application is discussed. A rapid, robust, and economically feasible GQDs with bright red fluorescence were obtained from *Mangifera indica*. The average particle size ranging from 2 nm to nearly 8 nm exhibits a combination of quantum confinement effect, excitation independent PL properties, and NIR emission ranging from 650 to 750 nm. Meanwhile, the preparation strategy and the precursors used here lead to tunable emission towards the NIR region and their potential use for *in vivo* bioimaging. Moreover, the temperature-sensitive fluorescence property demonstrated by these GQDs showed them to be an exceptionally stable temperature sensing probe. Highly biocompatible, novel, scalable, photostable, green synthesis mediated GQDs illustrated here are effectively applied for NIR imaging and nanothermometry applications that expand their key role in biomedical nanotechnology [66].

GQD materials were considered to be more advantageous than organic dyes in identifying multiplexed targets which is more effective in probing cellular imaging [67]. For example, sorted nitrogen-doped GQDs conjugated with polymers exhibit excitation wavelength independent PL spectra between the ranges of ultraviolet to NIR [68]. The high quantum yields and effective two-photon properties associated with these make them effective two-photon contrast agents. The successful utilization of these unique PL properties of GQD enables the simultaneous detection of four target cells at a single two-photon excitation wavelength.

4.7. Excitation-Dependent and Independent PL Properties of GQDs for Bioimaging Application

As we have seen, excitation-dependent PL property of GQD is a mysterious one. To engineer this excitation dependency, first the PL mechanism must be clarified. Only by resolving the PL mechanism, one can undeniably design truly tunable GQDs for a bioimaging purpose. An example of a GQD based system correlating the structural modulation, PL tunability and subsequent bioimaging application is shown in Figure 13 [69]. Here the tunability in PL properties is brought about by coating GQDs by polyethyleneimine (PEI) of two different molecular weights (PEI_{1800} & PEI_{600}). The yellow fluorescence of unmodified GQDs changed to blue and red fluorescence upon the modification with PEI_{1800} & PEI_{600}, respectively. Among the two systems, the GQD-PEI_{1800} depicts excitation-independent PL spectra whereas GQDs and GQD-PEI_{600} exhibits excitation-dependent PL spectra, which mainly arises due to their structural distinctions. In the GQD-PEI_{1800} system, PEI chains are attached to the GQD surface via electrostatic interactions. Since PEI is a well-known positively charged polymer with abundant amine groups it facilitates better electrostatic interactions with negatively charged surface of GQDs. It is affirmed that the blue shift in PL spectra of GQD-PEI_{1800} due to the conjugation of amine functionalities of PEI_{1800} with the π-π moieties of GQDs via amidation reaction facilitates electron transfer (ET) between them. Moreover, the conjugation domain is narrowed, which leads to the enlargement of bandgap. On the other hand, in the GQD-PEI_{600} system GQDs are placed as an encapsulated form on PEI_{600}. The red shift originated here is due to the enlargement of the core of GQDs (π-system) via cross-linking with PEI_{600}, also termed as cross-linking enhanced emission effect (CLEEE), which leads to the reduction in the bandgap of the GQD-PEI_{600} compared to that of the GQD-PEI_{1800}. Owing to this size-dependent luminescence, bandgap tunability of GQDs is achieved ensuing the multicolor *in vitro* imaging of human primary glioblastoma cell line 87 (U-87) U87 tumor cells [70].

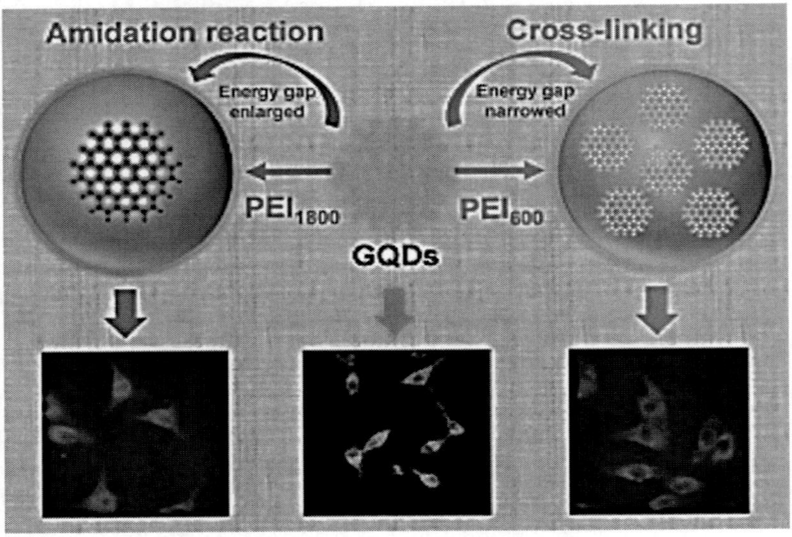

Figure 13. Scheme showing the polyethyleneimine-coated GQDs and the bioimaging application of PEI-GQD. Reproduced with permission from ref. [70].

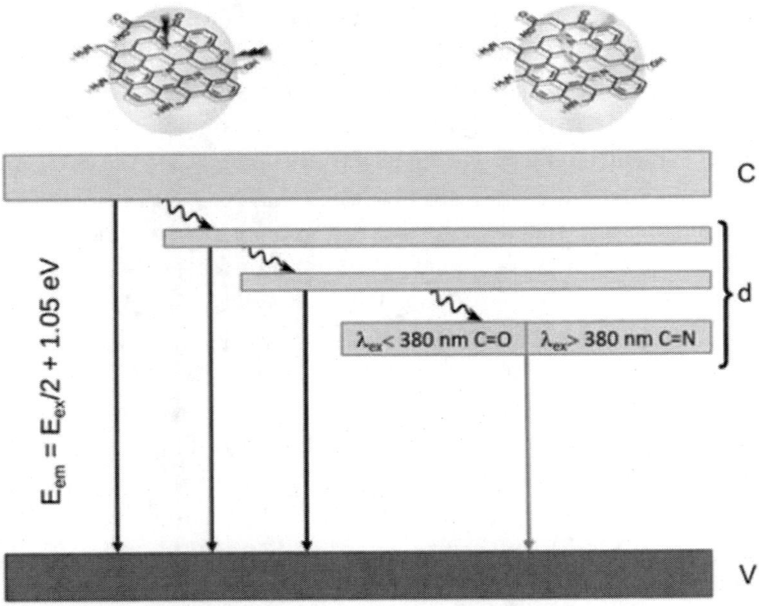

Figure 14. Proposed PL emission mechanism for the NUV−vis−NIR range by the localized $\pi^* \rightarrow \pi$ transition from the aromatic and sp^2 skeleton of N-GQDs and $\pi^* \rightarrow n$ transitions of C=O and C=N bonds. Reproduced with permission from ref. [10].

Determining the origin of emissive pathways of the PL mechanism is a critical factor in GQDs which influences their multicolor imaging properties. The following illustration (Figures 14 & 15) based on nitrogen-doped GQDs (N-GQDs) provides an insight into the various photophysical transition in GQD that ultimately decides the bioimaging capacity of the system. The microwave-assisted synthesized GQDs, in this case, show excitation-dependent broadband PL from the NUV-Vis-NIR range, due to the combination of band-to-band and defect-induced transitions. These broadband emissive GQDs were developed to provide high biocompatibility for the imaging of vascular smooth muscle cells (VSMC cells) [10].

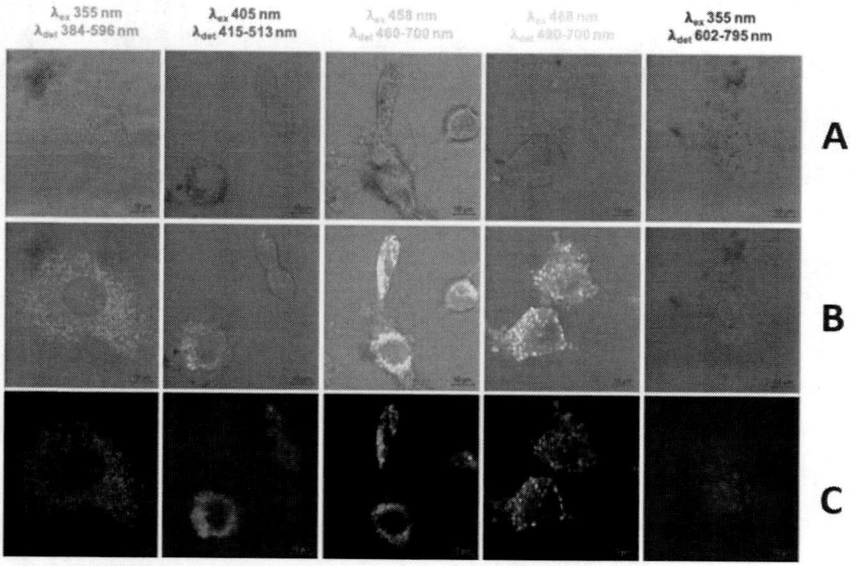

Figure 15. Confocal fluorescence images of VSMC cells incubated for 24 h with N-GQDs. Corresponding images under (A) bright, (B) fluorescence fields and (C) overlay. Reproduced with permission from ref. [10].

4.8. Monitoring the Cellular Dynamics Using GQDs

The above sections convey the correlation between the optical properties of GQDs and their bioimaging capacity in various cell lines. In addition to this, monitoring the cellular redox dynamics with the aid of bioimaging tool is also needed, which is still ambiguous, and it is the most advanced research area. A switchable fluorescent probe capable of visualizing the overall redox state and its corresponding dynamic changes concerning the various physiological event is at the forefront of the bioimaging technology. GQDs having the advantage of being small size and having fluorescence turn-OFF effect by means of photoinduced electron transfer (PET) from GQDs to various electron-withdrawing redox functionalities are known to be the most eminent candidate. As given in Figures 16, 17, and 18 the desferrioxamine-functionalized GQDs (GQD-DFOB) chelated with the redox Fe^{2+}/Fe^{3+} coupled probe are utilized for monitoring the triggered dynamic changes of the overall intracellular redox status states in HeLa cells. The PL associated with the GQD system is highly prone to PET process due to small size and quantum confinement effect. The redox couple, Fe^{2+}/Fe^{3+}, is the significant enzyme cofactor of all the living organisms. It is also shown that the GQD-DFOB-Fe^{2+} system is useful for detecting various intracellular reactive oxygen species (ROS), H_2O_2, and NO_x via fluorescence quenching although the GQD-DFOB-Fe^{3+} system is efficient for the detection of various antioxidants, ascorbic acid (AA), glutathione (GSH), and superoxide dismutase via fluorescence recovery strategy [71].

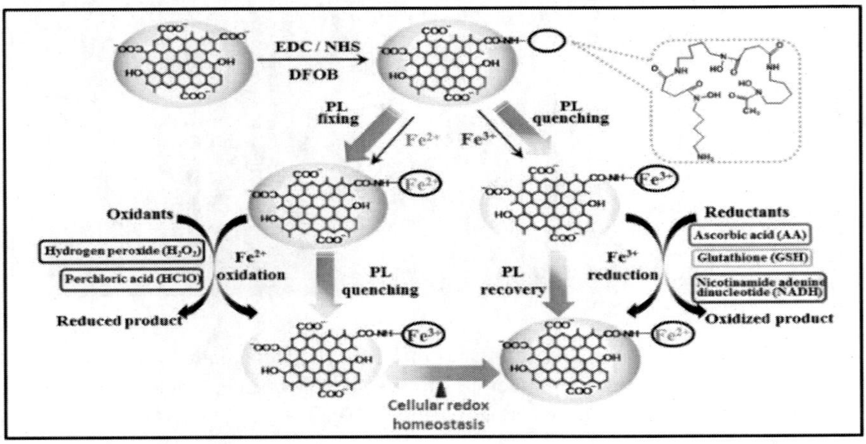

Figure 16. Schematic illustration of a redox-sensitive fluorescence-switchable probe based on a desferrioxamine-functionalized GQDs (GQD-DFOB) chelated with the redox Fe^{2+}/Fe^{3+} couple. Reproduced with permission from ref. [71].

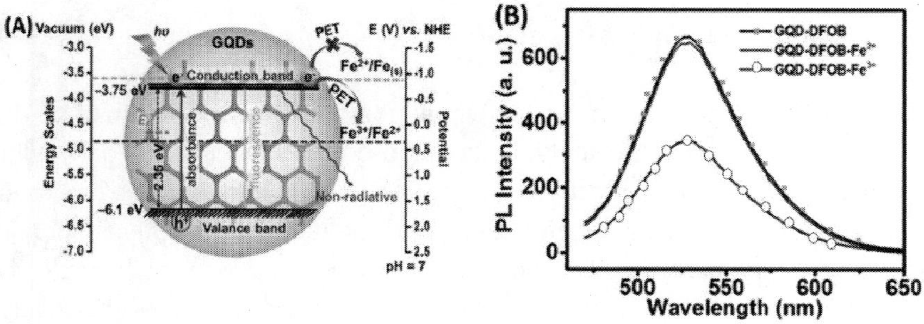

Figure 17. Diagrammatic illustration of the photoinduced electron transfer process between (A) GQD and the chelated Fe^{3+}. (B) PL spectra of GQD-DFOB chelated with Fe^{2+} or Fe^{3+} (0.1 mg/mL in PBS, pH 7.0). Reproduced with permission from ref. [71].

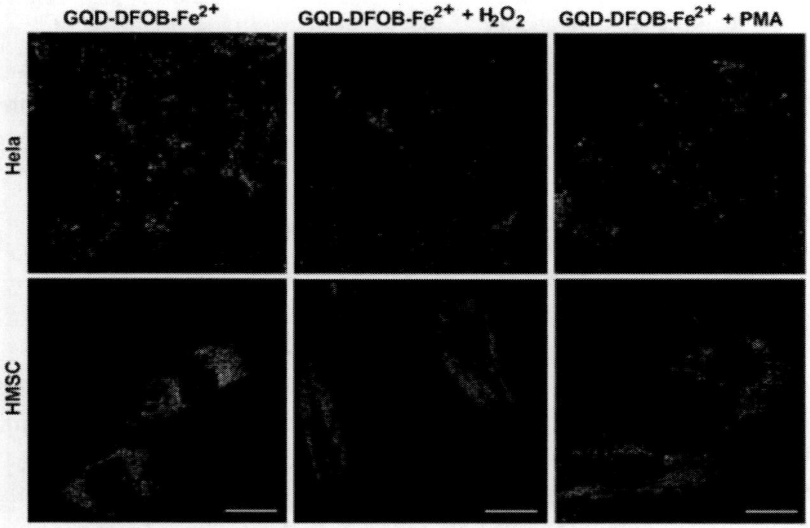

Figure 18. Confocal images of HeLa cells (top) and human mesenchymal stem cells (bottom) pre incubated for 1 h with 0.1 mg/mL GQD-DFOB-Fe^{2+}, in the absence (left) or presence of 10 μM H_2O_2 for 40 min (middle) or 10 μM PMA for 1 h (right). Scale bar = 10 μm. Reproduced with permission from ref. [71].

Briefly, a thriving perception of bioimaging application of GQDs, influence of the significant optical properties such as UCL, NIR emission, excitation dependent/independent emission on the bioimaging process from unmodified to functionalized GQD and intracellular redox dynamics provide numerous opportunities for evaluating the advanced bioimaging application of GQDs.

4.9. Biosensing Application of GQDs

Biosensors are defined as an analytical tool bearing a combination of a biological recognition element and the required transducer that is particularly connected with a data processing analyzer [72]. Generally, the biosensors integrate a biologically significant element through a physiochemical transducer which further produces a corresponding electronic signal. The electronic signal generated here is usually proportional to the analyte concentration and eventually transferred to a detector. Biosensors are generally classified based on the following facts [73].

- Nature of receptors involved in the biorecognition process, e.g., DNA, aptamer, enzymes, peptides, antibodies, molecularly imprinted polymer-derived sensors, and so forth.
- Nature of the transducer involved during the biorecognition process: optical (mainly fluorescence), electrochemical, photoelectrochemical, calorimetric, piezoelectric biosensors, and so forth.

One of the promising uses of the biosensing platforms is that they can fulfill the prerequisite of fast and specific real-time monitoring of biomolecules. Hence it is figured out that these sensing platforms can furnish ample chances for research and development in the biomedical field. Nanomaterials have been widely utilized for the fabrication of biosensors owing to their multifarious properties such as structural tunability, magnificent biocompatibility, strong absorption capacity, electro-catalytic performance, non-toxicity, long-term chemical stability, selectivity, sensitivity, repeatability, signal amplifying characteristics, photostability (against photobleaching and photoblinking), quite easy functionalization and so forth [74].

4.10. Optical Biosensors

Among the various biosensors developed, optical biosensors are the well-utilized promising systems for the detection of biomolecules. This technique is a combination of the light source, a modulating element, a biologically significant receptor, and a photodetector for interpreting the optical response of generated signal. The major optical transducers are listed as follows. All these techniques rely on the general principle of interpreting optical signals which are highly sensitive to the refractive index variation corresponding to the bio-recognition process [75].

- Fluorescence
- Photoluminescence (PL)
- Chemiluminescence (CL)
- Electrochemiluminescence (ECL)
- Fluorescence resonance energy transfer (FRET)
- Photoinduced electron transfer (PET)
- Aggregation-induced emission enhancement (AIEE)
- Surface plasmon resonance (SPR)

4.11. GQD-Based Fluorescence Biosensing Platforms

Fluorescence is defined as an event that occurs within a short lifetime from 10^{-9} to 10^{-8} seconds generated upon the absorption of higher energetic photons (shorter wavelength) and progressively emitting lower energy photons (longer wavelength). There have been numerous reports that demonstrate the in-depth details of the fluorescence biosensing applications of GQDs.

DNA, RNA, and aptamers are one of the primary targets of GQD-based biosensors [76]. The following example implies that pentaethylenehexamine and histidine-functionalized graphene quantum dots (PEHA-GQD-His) serve as the fluorescence biosensing scaffold for the microRNA (miRNA) detection. This detection is carried out through a molecular level beacon double cycle amplification strategy. The mechanism evolved here is quite complex, but it is really valuable. The specific binding of miRNA to the molecular beacon is the prime step that evokes the target and the molecular beacon cycle, subsequently producing a DNA nano assembly on the pentaethylenehexamine-GQD-histidine surface. Consequently, a turn-OFF behavior exhibits the probe through a photoinduced electron transfer mechanism between the probe and the analyte. The fast and high sensitivity detection was induced by this system towards miRNA in human serum samples. Within the linear range of 1×10^{-18} – 1×10^{-12} M this pentaethylenehexamine-GQD-histidine system depicts a 4.3×10^{-19} M limit of detection which signifies the potential use of GQD-based systems in biosensing application [77].

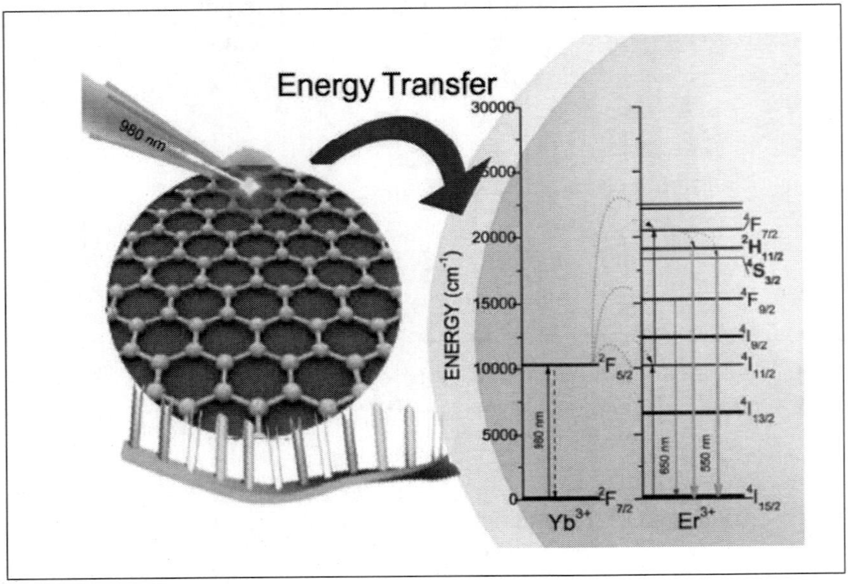

Figure 19. Proposed energy transfer mechanism between GQDs and UCNPs. (Reproduced with permission from https://pubs.acs.org/doi/10.1021/acsami.6b02361) [78].

Like covalent interactions, the non-covalent interactions also bring about a significant fluorescence response in GQD-based systems to be used in biomolecules detection. For example, a GQD-based biosensor integrated with single-stranded DNA (ssDNA), silicon dioxide (SiO_2), and up-conversion nanoparticles (UCNPs) act as a potential biosensing platform for the detection of miRNA sequences (Figure 19) [78]. The major mechanism involved, in this case, is the enhancement in up-conversion emission which occurs via the π-π stacking of GQDs on the ssDNA-UCNP/SiO_2 surface. Meanwhile, the association of ssDNA with their respective complementary miRNA sequences resists the up-conversion particles to react with GQD via π-π stacking that reduces the fluorescence emission intensity as a function of miRNA concentration. Here, with the aid of comparative emission/up-conversion measurements relative to a reference, the presence of complementary miRNA target sequences is detected with a limit of detection (LOD) of 10 fM [78].

Ratiometric fluorescence assay provides ultrasensitive detection of biomolecules with real-time monitoring, in which label-free detection provides a convenient method to monitor the small biomolecules, cells, and proteins with high sensitivity and selectivity for the high outturn drug discovery application. Moreover, these label-free detection methods are a good alternative for other reported procedures attributed due to their capability to furnish easy assay approaches to hit validation, un-altering the complex target groups, monitoring the interaction of drugs with cells, and evaluating the attraction of small molecule inhibitors to target proteins [79].

The ratiometric fluorescent probe based on GQDs described below adapted a signal amplification as well as a target recycling strategy for the DNA sensing procedure [80]. The signal amplification is designed to increase or amplify probe molecule (which in turn conjugated to the target nucleic acid sequence) induced signals. The major benefits of these methods are comprised of specific target detection, dynamic range, ease of use, reproducibility, selectivity, and sensitivity [81]. In the above-mentioned example, the assay system is designed in such a way that cutting the endonuclease enzyme assisted with the hemin DNAzyme/G-quadruplex biocatalyst via signal amplification strategy takes place. G-quadruplexes (G4) is defined as non-canonical nucleic acid form with stacked G-tetrads gathered by hydrogen bonds. The G-tetrads structure can bind hemin (iron (III)-protoporphyrin IX) to form a distinctive type of DNAzyme/RNAzyme with peroxidase-mimicking activity, which has been widely employed in multidisciplinary fields [82]. In the present example of the GQD system, o-phenylenediamine (OPD) acts as a hemin DNAzyme/G-quadruplex substrate. The oxidized product from this system (i.e., 2,3-diaminophenazine (DAP)) induces quenching of fluorescence intensity of GQDs (at 460 nm) along with the formation of a new peak at 564 nm. This implies the ratiometric behavior of GQDs that is utilized here for the sensitive and selective detection of specific DNA targets with a LOD of 30 fM.

GQD-based hybrid structures received lots of appreciation, especially for their biosensing applications. In earlier times the biosensing applications of GQD were limited to their pristine form only, later covalent functionalization, heteroatom doping, and some non-covalent approaches were adopted depending on the nature of the target analyte [83]. The following example illustrates the sensing of cholesterol (cholesterol being one of the significant molecules that take part in cell membrane designing, protecting cell structure and controlling normal metabolism) with the use of nitrogen-doped GQDs-chromium picolinate (CrPic) hybrids as the fluorescent probe. Nitrogen doping provides efficient surface grafting, improved fluorescence quantum yield, and superior biocompatibility which is the key requirement for an efficient biosensor. The implantation of chromium picolinate introduces a donor moiety in the probe whereas the GQDs are highly electron deficient, ultimately bringing about a donor-acceptor system. The fluorescence detection is achieved via a photoinduced electron transfer (PET) between chromium picolinate and N-GQDs. The chromium picolinate act as an efficient receptor for cholesterol detection, which is accomplished via binding affinity and π–π stacking between the CrPic/N–GQDs complex and cholesterol [84].

Ascorbic acid is yet another important analyte widely detected by fluorescence sensing strategy. Numerous studies showed the various capacities of the fluorescence quenching between the different valence states of metals, but the toxicity associated with these metals limits their wide application. A simple strategy with notable features based on GQD mediated system is illustrated here. It is worthy to mention that the fluorescent GQDs given in this example possess an ideal fluorescence emission peak at 540 nm. However, in the initial reports, GQDs exhibit emission peaks mostly less than 500 nm [85]. Moreover, this system has the advantages such as high selectivity, versatility, absence of heavy metal and so forth. The sensing process is established here via a principle defined as follows. It has already been reported that molecular systems that comprise benzoquinone structure are an excellent electron acceptor that can significantly quench the fluorescence intensity of conventional semiconductor quantum dots [86]. This simple strategy is successfully applied in the case of the GQD system which in turn is further applied for the detection of ascorbic acid. In the initial step, the phenolic hydroxyl groups of catechol are oxidized to benzoquinone. This is carried out in the presence of hydroxyl radicals produced by hydrogen peroxide (H_2O_2) under the catalysis of horse radish peroxidase (HRP) which leads to fluorescence quenching of GQDs. Meanwhile, during the second stage, the H_2O_2 produced is consumed by the ascorbic acid, which hinders the formation of o-benzoquinone resulting in the fluorescence recovery. This metal-free strategy based on GQD structures provides potential application in the pharmaceutical field [87].

Another class of biosensors is magnetic fluorescent biosensors. The magnetic particles are used to modify the fluorophore molecules by which the fluorescence properties are altered. In earlier times GQDs were modified with these kinds of magnetic metal particles to decrease the magnitude of surface/trap states, but most recently it was identified that the modification with magnetic particles not only removes the trap states but can also improve the optical sensing capacity of GQDs. In the following example, a turn-on magnetic fluorescent biosensor based on GQD is highlighted [87]. Here the electrochemically synthesized GQDs are modified with a magnetic agent Fe_3O_4 and epithelial cell adhesion molecules which act as aptamer molecules. The whole system is finally coupled with MoS_2 nano sheets that ultimately result in a nanohybrid system with GQDs. This fused structure offers turn-on biosensing behavior towards circulating tumor cells. Moreover, this biosensor could quickly sense and label circulating tumor cells within 15 min, defeating certain other steps of detection techniques. This approach could detect up to 10 tumor cells in human blood serum with a linear detection range and detection limit of 2–64 nM and 1.19 nM, respectively [88].

4.12. Fluorescent Immunoassay

Fluorescent immunoassay is a versatile immunoassay method that is a widely applied technique for the simple, rapid, and sensitive detection of numerous compounds including drugs, hormones, proteins and so forth. This strategy has been utilized in various areas for the *in vitro* diagnostics (IVD) assay development [89]. The major principle involved in this technique is the biochemical detection of antigens or analyte molecules using antibody-labelled fluorophore. This technique possesses numerous advantages such as rapid detection of the analyte with high sensitivity and specificity, simplified assay design, and easy fabrication of point-of-care devices. The implementation of the fluorescent-based immunoassay technique at the point-of-care application is widely exploited. Recently, fluorescent immunoassay acted as a detection probe or reagent which favors the implementation of high sensitivity immunoassay system [90]. This includes the narrow wavelength with low-cost light sources, novel stable fluorophores with very wide Stokes shifts, stable solid-state light detectors, and microprocessors to process and analyze the data from each test. When a fluorescent detection system is connected to a lateral flow assay having a powerful yet low-cost analyzer, it leads to improving the assay performance and avoiding the misconception generally associated with visually-read point-of-care assays [91, 92].

Consider the following example, in which a brand new dual-mode immunoassay protocol is illustrated. The combined effect of both surface enhanced resonance scattering (SERS) and fluorescence phenomena is utilized here for the fabrication of this immunoassay [93]. Surface enhanced Raman spectroscopy (SERS) is a dominant technique involving the principles of vibrational spectroscopy methods that permits exceptionally sensitive structural analysis of ultra-small concentration analytes along with the amplification of electromagnetic fields created by the excitation of localized surface plasmons [94]. In the above-mentioned example, the water-soluble GQDs served as a fluorescent indicator and a source of Raman signal. The collective properties of SERS and fluorescence signals are used here to monitor the immunoreaction effects of tuberculosis (TB) antigen, CFP-10. The high sensitivity and selectivity are the key factors that determine the efficiency of fluorescent immunoassay probe and in this particular case the dual-mode of analysis results in high sensitivity and selectivity with a detection limit of 0.0511 fg mL^{-1}. Besides that, this dual-mode immunoassay recognition on a solid scaffold further leads to high sensitivity and admirable multiplexing capacity.

Fluorescence resonance energy transfer (FRET) based fluorescence immunoassays are yet another promising approach widely accepted. FRET is a sensitive and authentic method used in fluorescence analysis [95]. Fluorescence resonance energy transfer is defined as a distance-dependent physical event by which energy is transferred non-radiatively from an excited state fluorophore, (that is the donor) to another fluorophore called acceptor. This energy transfer mainly occurs by means of intermolecular long-range dipole-dipole coupling interactions. FRET process is known to be highly efficient when it meets the criteria of molecular proximity distance of 10–100 Å. In other words, if the donor and acceptor are situated within the Förster radius (typically 3–6 nm) half the excitation energy of the donor is transferred to the acceptor) [96].

The use of FRET-mediated GQD-based fluorescence immunoassay is illustrated below. Von Willebrand factor (vWF) is a multimeric plasma glycoprotein detected by means of graphene quantum dots (GQDs) and core-shell Ag@Au nanoparticles (Ag@Au NPs) nanohybrids, which act as a donor-acceptor system involving an efficient FRET process. The use of Ag@Au core-shell type structure not only includes the features of a single metal element but also has more efficient performance than a single-metal element. Further, it improves the stability and dispersion of the GQDs and enhances the biocompatibility of the whole material. FRET process between GQDs and Ag@Au NPs led to the fluorescence diminishment of the former one. Meanwhile, the immunoreaction between vWF and Ag@Au-Ab nanocrystals resulted in the quenching of FRET efficiency and the magnitude of fluorescence recovery of GQDs. Apart from the covalent interactions, the non-covalent linkages of GQDs in this example lead to efficient FRET pair. Structural features of GQD depict that it is easy to modulate GQDs as a donor or acceptor type of moiety depending upon the required applications [97].

A similar type of complex system involving GQDs and Ag@AuNPs pairs given below utilizes Ag@AuNPs as the acceptors to quench the fluorescence of GQDs through the FRET process. In this example SARS-CoV-2 (2019-

nCoV) antigen (its diagnosis received widespread attention recently due to the reason being the cause for an acute infectious disease) detected by means of GQDs and Ag@AuNPs pair provides the analysis in the range 0.1 pg mL^{-1} – 10 ng mL^{-1}, and the limit of detection 50 pg mL^{-1}. The mechanism attributed is as follows: in the presence of 2019-nCoV mAb, a steric hindrance was noticed among the Ag@AuNPs-NCP (2019-nCoV antigen) complex and GQDs, which decreases the FRET efficiency and reimposes the fluorescence of GQDs. Moreover, here the GQDs nanohybrids coupled with antigen by means of electrostatic interactions without any modification and simple procedures reveal a novel platform for the detection of clinically relevant antigens [98].

Apart from the non-covalent strategies, the covalently functionalized GQDs systems will also pave a path for efficient immunoassay analysis. Covalent functionalization of GQDs renders superior optoelectronic properties to GQDs. GQDs usually contain functional groups like carboxyl, hydroxyl, carbonyl, and epoxide at their edges that can act as reaction sites for further functionalization and alter PL emission from the dots by changing their electron density. Recently, many efforts were put forward in order to functionalize the GQDs with suitable molecular systems such as -OH, -NH_2, -SH containing conjugating molecules, conducting polymers, macromolecules (e.g.: porphyrin) etc. Surface chemical functionalization has been proven to be a successful approach for exploiting the photoluminescence properties of GQDs. The real goal of this approach is an effective charge transfer between the GQDs and functional groups that can tune the band gap of GQDs and along with this, it leads to different photoluminescence properties. In comparison with the unfunctionalized GQDs, the functionalized GQDs can effectively tune the band gap properties, biocompatibility, optoelectronic properties and so forth [14, 15].

Consider the following case as an example (Figure 20) that demonstrates the designing and optimization of an immunosensor based on amine-functionalized GQD-methyl orange fluorophore quencher pair for ferritin estimation. In this example, amine functionalization significantly increases several optoelectronic properties such as quantum yield, stable photoluminescence, Stoke shift, molar extinction coefficient and so forth. High fluorescent properties of these amino functionalized GQDs are exploited by conjugation with anti-ferritin antibodies (Ab) for specific fluorescence-based immunosensing of ferritin. One advantage is being a sensitive and selective method for the ferritin sensing for a wide linear range with no interference from other foreign species. Moreover, the ferritin-specific antibodies used here results in high selectivity for the proposed sensor. The high selectivity obtained here is achieved through the incorporation of ferritin-specific antibodies. There are several reports for the detection of ferritin based on GQDs but in comparison to that these GQD functionalized systems are obtained through a quite simple strategy and also GQDs can be easily modified by substituting the antibody for the desired analytes expanding its applicability in real-time monitoring [99, 100].

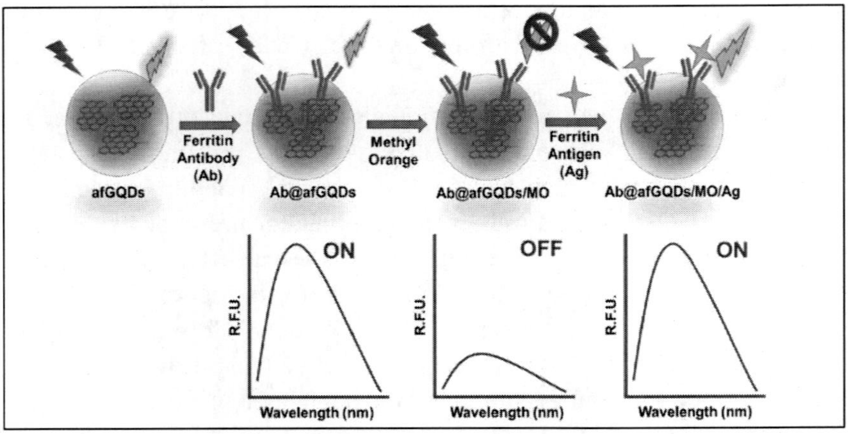

Figure 20. Plausible mechanism for ferritin sensing using Ab@afGQDs and MO as the fluorophore−quencher couple[a]. [a]afGQDs, Ab, Ag, and MO refer to amine-functionalized graphene quantum dots, antibody, antigen, and methyl orange, respectively. R.F.U.: relative fluorescence units. Reproduced with permission from ref. [100].

5. Energy-Related Applications

The demand for energy supply is progressively raised due to the hike in population. The outsmart way to meet this need is renewable energy resources, among which solar energy is a magnificent choice. Solar cells are the pre-eminent one which have the power of efficacious conversion of light energy to electrical energy. The graphene quantum dots (GQDs) find possible applications in the field of energy such as light emitting diodes (LEDs) [101], supercapacitors [102], fuel cells [103], and lithium-ion batteries [104], other than the photovoltaic devices [105]. The characteristic quantum confinement and edge effect of GQDs paved their way to these applications. In the upcoming topics the application of GQD in photovoltaics and light emitting diodes (LEDs) are discussed.

5.1. Solar Cells

Fundamentally solar cell is a semiconductor since it is made up of n-type and p-type semiconductors which can carry electrons from one end to other [105]. The n-type semiconductor is the negative electrode due to the excess of electrons and p-type is a positive electrode due to the presence of excess holes. When these two electrodes are in contact with each other, there generate a depletion layer in which the electrons are fused with holes, as shown in Figure 21 [106]. In this depletion zone the charges are get reversed (holes are seen in n-type and electrons in p-type) which generates an internal electric field that prevents further movement of electrons from n-type to p-type layer. Upon the fall of sunlight over the cell, the electrons from the materials get excited and the internal electric field will move electrons back to n-type and holes to p-type layers. If they are connected to an external circuit electricity will be generated due to the flow of electrons and holes [106].

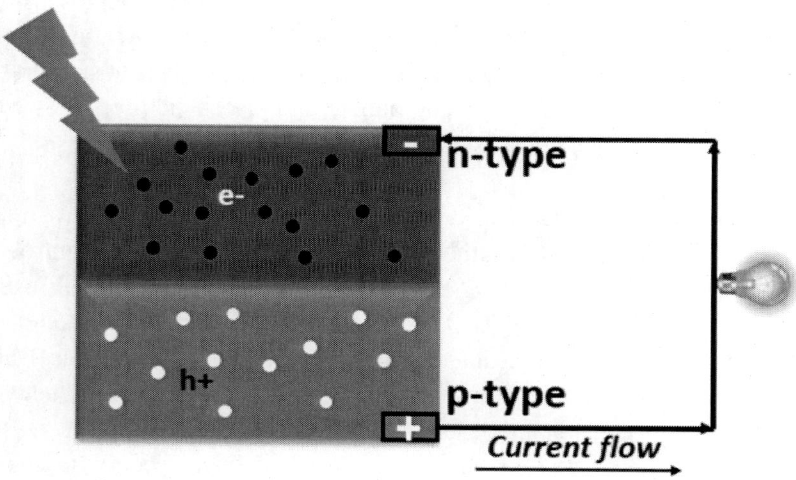

Figure 21. Diagrammatic representation of typical solar cells [106].

There exists manifold of a solar cell, namely dye-sensitized solar cells (DSSCs), organic solar cells (OSCs), bulk heterojunction solar cells (BHSCs), tandem solar cells (TSCs) and so on [105, 107-109]. All the above follow diverse charge transfer mechanisms and possess different kinds of photoactive materials and methods of fabrication, although the basic principle remains the same as elucidated above. Herein we are focused on the application of GQDs in the above-mentioned solar cells. Tunable bandgap, edge effect, good carrier transport mobility, and high surface area are the salient features of graphene quantum dots that makes them have possible applications in the field of photovoltaic devices [108].

5.2. Dye-Sensitized Solar Cells (DSSCs)

Out of a number of solar cells, dye-sensitized solar cell still receives much attention due to its characteristic features such as cost effectiveness, easy fabrication, and so on [110]. GQDs find possible applications in the photoanode, photosensitizer, counter electrode and in solid state electrolyte in dye-sensitized solar cells (DSSCs). Among of these, the role of GQD as photosensitizers, especially as a co-sensitizer, becomes the prior one, due to the light harvesting ability, conductivity, and presence of anchoring groups on the edges of GQDs [111]. A detailed investigation on the application of GQD in the four components of DSSCs such as, photoanode, photosensitizer, counter electrode and electrolyte are also included as follows. Among the copious properties of GQDs, QCE and edge effect are the best to support their use in DSSCs. They produce unique optical and electronic properties which can augment the power conversion efficiency (PCE) of solar cells.

5.2.1. GQDs as Photoanodes

The pertinent part of DSSCs is the photoanode, in the wake of its control over the photocurrent and photovoltage of DSSCs. Conventional materials used as photoanodes are TiO_2, SnO_2, ZnO and so forth [105]. High bandgap and good conductivity of these metal oxides endorses their role as photoanodes. But the photoexcited electrons while moving through these nanostructured networks encounter a lot of nanoparticle boundaries, which leads to the decrease of charge transfer and increased recombination with dye and electrolyte. In order to avoid this issue, the coalescence of other materials like the highly conducting carbon materials into TiO_2 nanostructures have been practiced [112, 113]. Amongst carbon nanomaterials, the exceptional properties of GQDs added impetus to decorate the TiO_2 nanofibers (NFs) and use as a photoanode [114]. Presence of GQDs over TiO_2 NFs escalates the dye adsorption thereby generating more photoelectrons and faster charge transfer.

The good conducting GQDs themselves reduce the back recombination. These factors favor the short circuit current (Jsc), fill factor (FF), and finally the PCE of the DSSCs. GQDs may get varied according to their mode of formation and they have different PCE when incorporated with TiO_2. Binder free TiO_2 paste was prepared to use in DSSC anodes [115].

GQDs are viable hot carrier (electron/hole) chromophores that exceed the Shockley-Queisser limit and can theoretically achieve a PCE near to 66% for a solar cell [116]. Charge transfer dynamics from GQDs to TiO_2 was determined by using femtosecond time resolved second harmonic generation and it was perceived that in the ultrafast time scale of less than 15 femtoseconds the photoexcited electrons of GQDs get transferred to the conduction band of TiO_2 [116]. As the hot electron lifetime of GQD is about hundreds of picoseconds and electron transfer from GQDs to TiO_2 interface takes less than 15 femtoseconds, this leads to the aberrant stability of the cells for more than 30 days [114].

Another photoanode of GQD-TiO_2 procured an efficiency of about 6.17% according to a size dependent study, in which maximum efficiency was attained by the GQDs having 5 nm size [117]. Over and above, GQDs act as a light scattering layer on top of the anodic TiO_2, and clusters of this spherical nanoparticles enhance the rate of light scattering. Along with that, GQDs introduce new defect states/trap states that reduce the electron-hole recombination [118]. Significantly, it was noted that the GQD coated TiO_2 electrodes require low dye concentration to result in higher efficiency [119]. Similarly, a polyaniline (PANI)-GQD hybrid was found to be an active substituent for TiO_2 anodes. Doping of carboxylate anchored GQDs to the PANI chains results in uncoiling and elongation of the PANI chains. The highly conducting PANI receives the excited electrons from dye molecules and passes to the LUMO of GQDs to the external circuit [120].

5.2.2. GQDs as Photosensitizers

Basic criteria to use a material as a photosensitizer is its light harvesting ability and effective binding with TiO_2. The GQDs possess a good light absorption and also, the presence of carboxylic (-COOH) and hydroxyl (-OH) groups on the edges of GQDs favors their binding with TiO_2 (Figure 22) [121]. Moreover, their nontoxicity, tunable bandgaps, and high mobilities proclaims their potential as light harvesting materials in photovoltaic devices.

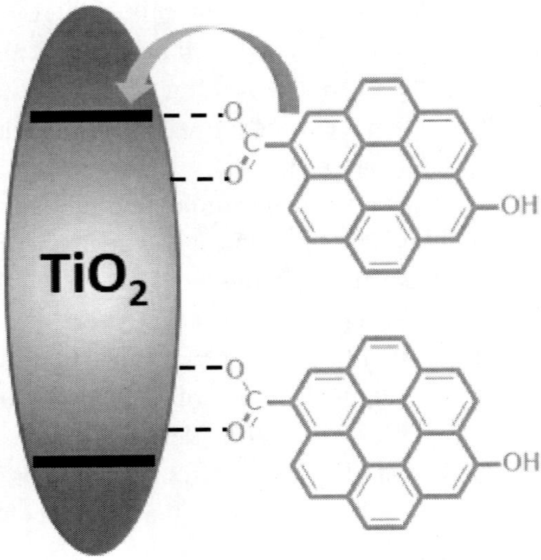

Figure 22. Schematic representation of bonding between TiO_2 and GQDs.

Use of solely GQDs as a photosensitizer do not offers a guaranteed efficiency, mainly because of their UV range absorption. But their higher open circuit voltage agrees with their charge carrier separation, mobility, and collection [122]. Other than single dye usage there is another stream of photosensitizers in DSSC, i.e., co-sensitized dyes. Co-sensitization strategy has ample potential in DSSCs. In this strategy two or more dyes with complementary absorption are used as photosensitizers, which results in a panchromatic spectrum of absorption [123]. Thus, the paramount region of solar light is absorbed by the cell that leads to higher power conversion efficiency. It was observed that more studies were based on co-sensitized GQDs rather than a single GQD dye [124]. The major drawback of DSSCs is the aggregation of dyes over the semiconductor, which is eschewed by these quantum dots [121]. Comprehensive representation of cascade of electrons through the co-sensitized dyes is as shown in Figure 23 [34, 35, 113].

The electron transfer within a GQD based DSSC follows either a charge transfer or an energy transfer mechanism or both together [122]. Energy level diagram in Figure 23 demonstrates the feasibility of charge transfer mechanism, while the overlapping spectrum between the emission of GQDs and absorption of other dye molecules narrates the energy transfer [124] within the systems.

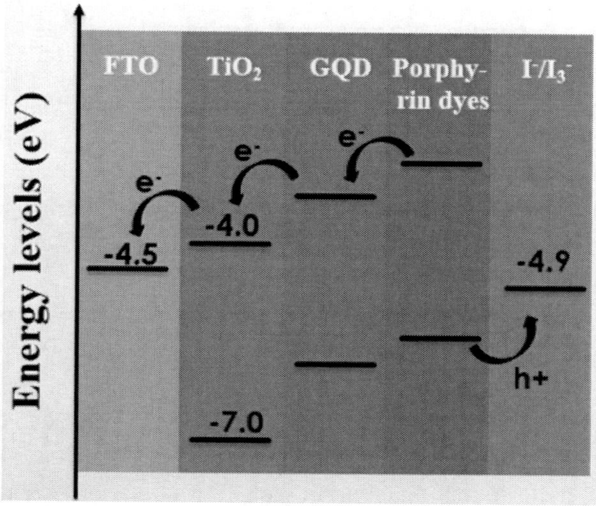

Figure 23. Energy level diagram of GQD based hybrid sensitizers in DSSCs.

GQDs have enough potential to replace ruthenium-based dyes in DSSCs with ZnO photoanode, because the spectral absorption range of ZnO dipped in GQDs is wider than ZnO in N719 dye and possess a significant reduction in the resistance for the electron transfer at the ZnO/dye/electrolyte interface, for the prior [125]. Likewise, Prabhagar et al. addressed some of the snags of DSSCs by using boron-doped GQDs (B-GQDs) as co-sensitizers and ZnO as the photoanode. The high surface area and extended optical pathlength of B-GQDs enhance the concentration of dye loading and light harvesting. Also, the high charge carrier mobility of these quantum dots improves the power conversion efficiency of DSSCs [126].

Next are Forster resonance energy transfer (FRET) based DSSCs, in which the FRET acts as a reason for electron cascade within the solar cells [124]. In the hybrid systems GQDs act as energy donors and N719/N3 as acceptors [122, 124], due to an effective overlapping between the fluorescence spectrum of GQDs and absorption spectrum of the standard ruthenium (Ru) dyes. Besides that, the excitation dependence of photoluminescence and adequate quantum yield (QY) have ample role in the energy transfer. When GQD co-sensitizes other dyes three possible mechanisms favor the efficiency such as (i) an augmented charge separation and mobility due to the continuous alignment of energy levels, (ii) feasibility of energy transfer from GQD to Ru dyes, and (iii) presence of GQD inhibits the back electron transfer from dye to the redox couple [122].

Excitation-dependent photoluminescence and up-conversion luminescence of GQD are like the two sides of a coin. Studies showed that the GQDs with up-conversion luminescence nature are better materials in FRET based photovoltaics. This property is caused by the multiple photons activation in the GQD system. The photovoltaic efficiency of DSSCs fabricated from these GQDs along with D719 dye was obtained as 7.95%, due to the greater light absorption of D719 via the FRET and up-conversion nature of GQDs [127].

It was noted that the GQDs can improve the efficiency of natural dyes as well [128]. Electronic energy levels of the dyes get modified by the presence of GQDs. Predominantly it is spotted in the absorption and emission spectra of the dyes. In a DSSC, the fluorescence property of a dye can be considered as an efficient electron-hole recombination. Thus, decreases in emission intensity can indicate the reduction in their recombination. Insertion of GQDs to the natural dyes, such as Ulva and Gracilaria, lowers the photoluminescence intensity and thereby yields a higher efficiency [128].

In short, GQDs can be effectively added to a DSSC as a co-sensitizer, but it does not offer much higher efficiencies than other fluorophores or semiconducting quantum dots, by dint of low quantum yield and finite spectral range of absorption [122].

5.2.3. GQDs as Electrolytes

The application of GQDs as an electrolyte is a flabbergasting one, owing to the odd behavior of electrolyte in comparison with anode, cathode, and dye. Liquid electrolytes like ionic liquids are the novel ones. The leakage and volatilization are the serious hiccups in their way to be used in DSSCs, therefore, scientific community started looking at solid state electrolytes. In this context the thirst for electrolyte substitutes increases and now it reaches at gel electrolytes, solid polymers, inorganic nanoparticles and so forth [129]. Nowadays carbon-based materials are used as electrolytes, yet only limited attention is acquired by GQDs.

An efficient quasi-solid-state electrolyte of GQD covalently functionalized with imidazolium based ionic liquid (GQDs-IL) acquired an efficiency of 4.57%, which was about 2.23% for GQD alone as an electrolyte in DSSCs [130]. Resistance of charge transport at the TiO_2/GQDs-IL interface is small, thereby enhancing the electron transfer from cathode to anode and promoting dye regeneration. In effect, the complete ionic transport through the electrolyte gets augmented. Besides the twigs of GQDs-IL binds strongly with

TiO_2 and protects the dyed TiO_2 from dye degradation that results in the improved stability of the cells.

5.2.4. GQDs as Counter Electrodes

The leading role of counter electrode in the work of DSSCs is that they act as catalyst for the reduction of electrolyte (commonly I_3^-). Best ever counter electrode up to this time is noble platinum (Pt) metal, even though high cost and low storage limits its practical applicability [131]. In this context the development of low cost, binder free counter electrodes is highly demanding and leads our focus towards the promising materials called GQDs. Them having the quantum dot sizes results in a greater number of edge carbon atoms, which are highly energetic and activate the catalysis of I^-/I_3^- electrolyte couple. Moreover, this property of GQDs vanquishes the chances of deactivation pathways, leading to better efficiency and stability of a DSSC with GQDs as a redox catalyst [132]. Together with that the excessive edge defective states, that are present on edge planes (both planes of GQDs are differentiated in Figure 24) of 0D GQDs augment its catalytic activity [133].

GQDs authenticate their role as counter electrodes, albeit the hybrid counter electrodes with GQDs seized more recognition [135]. For instance, GQD decorated carbon aerogels (GQD/CA) are one among them [132]. By the same token, GQD with strontium ruthenate ($SrRuO_3$) achieves a higher efficiency of about 8.05% (Table 1) [136]. Good conducting GQDs favor the connection between $SrRuO_3$ particles and generate a conductive network for this counter electrode. Synergic effect of both components enhances the charge collection and electron transfer, mainly provided by the sp^2 network of GQDs [136]. GQDs modifications by polypyrrole (PPy) and cobalt sulfide (CoS) gives augmented PCE of 5.27% and 7.30%, respectively (Table 1). The GQD doping by PPy engenders more porous structures on densely structured PPy, increased active sites, and higher catalytic efficiency towards the electrolyte [137]. Meanwhile, the negatively charged GQDs adsorb Co(II) ions and tailor the growth of CoS nanosheets, when applied as counter electrode in DSSC showing superior PCE than Pt and CoS due to the synergic effect of CoS and GQD [135].

Table 1. Overview of the efficiency of dye-sensitized solar cells by using GQDs as photoanode, photosensitizer, electrolyte, and counter electrode

Metal Oxide	GQD Derivatives	Photosensitizer	Efficiency (%)	Reference
TiO_2	GQD as co-sensitizer	N3	2.15	[122]
TiO_2	GQD a co-sensitizer	N719	7.96	[124]
TiO_2	GQDs-TiO_2 nanofibers	N719	6.22	[114]
TiO_2	GQD doped polypyrrole as counter electrode	N719	5.27	[137]
TiO_2	CoS-GQDs as counter electrode	N719	7.30	[135]
TiO_2	GQDs functionalized with imidazolium based ionic liquid	N719	4.58	[130]
TiO_2	GQD as electrolyte	N719	2.23	[130]
TiO_2	$SrRuO_3$ decorated with GQDs as counter electrode	N719	8.05	[136]
-	Polyaniline-GQD hydrids as photoanode	N719	3.12	[120]
TiO_2	GQD-TiO_2 as photoanode	N719	6.17	[117]
TiO_2	GQDs@graphene foam nanosheets as counter electrode	N719	9.59	[133]
TiO_2	GQDs as co-sensitizers	Natural dyes Gracilaria and Ulva	0.73 & 0.75	[128]
TiO_2	N,F,S-codoped GQDs as a part of photoanode	N719	11.7	[138]
TiO_2	PANI-GQDs nanocomposite as counter electrode	N719	1.60	[139]
TiO_2	GQDs modified TiO_2 photoanodes	N719	6.10	[119]
TiO_2	Single layer GQDs as co-sensitizer	N719	8.92	[140]
TiO_2	GQD as a light scattering layer over TiO_2	N719	5.01	[118]
ZnO	ZNSs/GQDs composite as anode	N719	5.27	[141]
ZnO	B-GQD as co-sensitizer	N719	3.7	[126]
ZnO	-	GQD	1.26	[125]

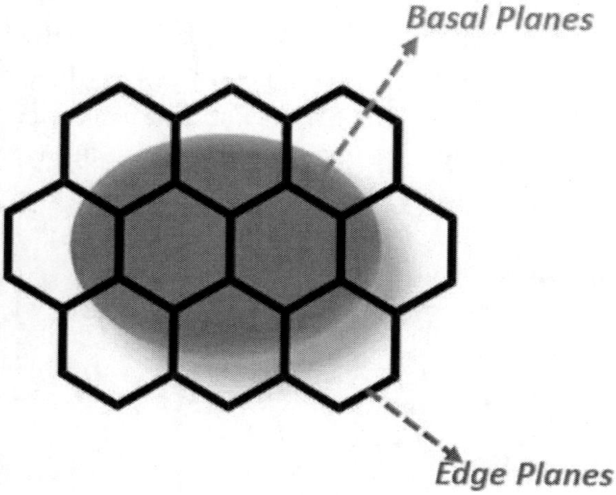

Figure 24. Schematic diagram of basal and edge planes of Graphene Quantum Dots (GQDs) [134].

5.2.5. Organic Solar Cells (OSCs)

Organic solar cells are a category of photovoltaics that uses conductive organic polymers or small organic molecules to assimilate light and carry over the charges. Sporadically they are called as polymer solar cells or plastic solar cells. Modulations in organic molecules are comparatively easy and cheap, resulting in potentially low-cost photovoltaic devices. Besides, these molecules show higher molar absorption coefficients than inorganic solar cells. Therefore, OSCs possess low efficiency, stability, and strength in comparison with silicon cells [142].

A variety of junctions are present in organic solar cells, in which a few are discussed below and shown as Figure 25 [143].

Junction types:

- Single layer - This is the simplest form of organic photovoltaic (PV) devices. An organic active material of donor or acceptor type is sandwiched between two metallic electrodes.
- Bilayer - As the name stipulates, this contains two organic layers with contrasting electronic behavior is stacked between two conductive electrodes. Junctions of this kind are also known as planar donor-acceptor heterojunctions.
- Discrete heterojunction - Three-layer junction like two acceptor and one donor or vice versa generates higher efficiencies.

- Bulk heterojunction - In contrast to the stacking arrangements, active organic layer of bulk heterojunction possess blend of donor and acceptor materials. Thus, the thickness of blend can be made larger for effective photon absorption. Variations of this type are known as graded heterojunction and continuous junction.

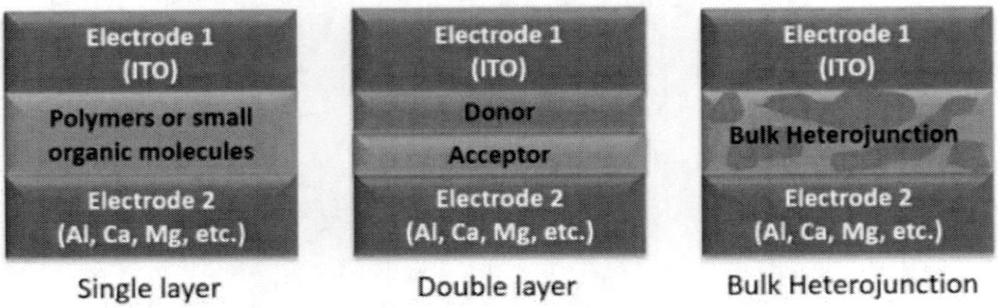

Figure 25. Schematic representation of different junctions in an organic solar cell [143].

To date, research is focused on cost-effective organic photovoltaic materials, namely, organic semiconductor quantum dots [144]. Frequently used polymers in OSCs are poly(3,4-ethylenedioxythiophene) (PEDOT), poly(3-hexylthiophene) (P3HT), polyethylene terephthalate (PET), poly(styrenesulfonate) (PSS), phenyl C61 butyric acid methyl ester (PCBM), and so on [145]. The HOMO-LUMO levels of GQDs are in between those of polymers and the metals which agrees with their applications in OSCs [146]. Subjected to the number of components present in the active layer of OSCs, they are categorized as binary, ternary, quaternary etc. [144, 147]. Combining a greater number of materials has the advantage of good complementary absorptions, as well as the synergistic interactions among them. The surface morphology of active layer is affected by the number, type, and ratio of donor/acceptor materials.

Identical to the case of DSSCs, complete replacement of electron acceptor and donor systems in the active layer of OSCs by GQDs results in lower performance of DSSCs [144, 147]. Ergo, the combination of active materials and GQDs is under scrutiny. Moving to the electron accepting role of GQDs in OSCs instead of efficient acceptors like PCBM, they acquired a PCE of 1.28% [144]. While aniline functionalized GQDs (ANI-GQDs) along with P3HT, i.e., P3HT:ANI-GQDs have 1.14%. Excited state charge transfer is reckoned as the mechanism at the P3HT:ANI-GQDs interface, in which excitons relax via a non-radiative relaxation process [147].

Limited number of studies were found on the differences between the role of GQDs on the binary and ternary hybrid OSCs. A binary system, P3HT:GQDs, and ternary system, P3HT:PCBM:GQDs, were compared and it was perceived that GQDs act as electron acceptors in both cases [148]. Besides, their increasing concentration results in agglomeration which acts as recombination sites in binary system and inhibits uniform mixing of P3HT and PCBM that leads to lowering of charge separation and transport in ternary system. Fullerene structure of PCBM easily adsorbs GQDs via π-π interactions/van der Waals forces and results in efficient exciton dissociation interfaces, which all together contributes to higher efficiency in ternary with respect to binary [148].

Also, the role of GQDs as a hole transport layer (HTL) has been corroborated by certain scientific reports. Low conductivity and non-uniform structure are the impediments faced by graphene oxide (GO) as a HTL. Meanwhile the more uniform structure and tunable bandgaps of GQDs are the steppingstones for their HTL application. Such type of cells exhibits longer lifetime and better reproducibility [149].

Literally, the bulk heterojunction solar cells, which are named characteristic to their cell structure, are the subcategories of well-known solar cells (SCs) such as silicon based SCs, DSSCs, organic SCs and so forth, which is encompassed in oncoming part of this chapter.

5.2.6. Bulk Heterojunction Solar Cells (BHJSCs)

Heterojunction is a concept of two or more materials with different electrical properties kept in contact with each other [150]. It is somewhat like a donor-acceptor (D-A) system in OSCs or a p-n junction in inorganic solar cells. The presence of light engenders excitons in the donor molecules and these charges move towards the electrodes via the organic layers [150]. As mentioned earlier a bulk heterojunction is produced from the interpenetrating networks of donor and acceptor modules sandwiched between the cathode and anode electrodes. This will create a larger interface and shorter diffusion path length, favoring enhanced efficiency.

Three types of band alignments are possible for a semiconductor heterojunction solar cell, namely straddling (type I), staggered (type II) and broken (type III), which are depicted as Figure 26 [151].

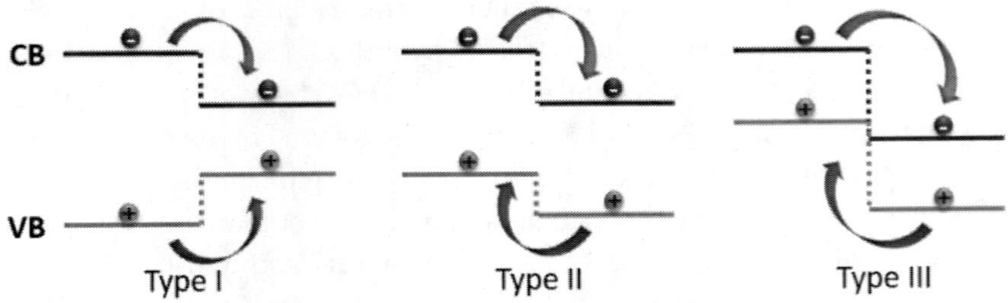

Figure 26. Possible band alignments in a heterojunction solar cell.

Graphene and its novel derivatives like GQDs find wide applications in heterojunction solar cells due to their large conductivity, optical transparency, good mechanical properties, and magnificent physico-chemical stability [2]. 2D graphene sheets undergo fast aggregation due to poor dispersion in common solvents which is the main hurdle faced for their application in devices. In this situation, GQDs become more important. Organic/inorganic hybrid solar cells are the enthralling one among BHJ solar cells, in which integrating one more layer has been enhancing its performance. Such a layer has many advantages like providing a panchromatic spectrum for the cell, hindering the back contact of excited electrons there by reducing recombination, and augmenting the donor-acceptor reactions [152]. The properties of GQDs as mentioned earlier, such as cost effectiveness, being eco-friendly, and stability endorse their usage in BHJs compared to current organic materials. Over and above that the blending of other polymers or functionalization can enhance the efficiency of GQDs based solar cells focusing on the electron accepting nature of GQDs. As an example, a binary blend of GQDs with P3HT polymer shows an increase in their PCE because of the increased surface area of P3HT:GQDs for p-n interface generation and augmented electron mobility [144].

A cell in which GQDs is a ternary blend between conventional inorganic and organic materials, i.e., TiO_2/GQDs/P3HT procured an efficiency of 3.16% that was 2.04% initially without GQDs [152]. Also, a higher degree of charge transfer was observed in TiO_2/GQDs/P3HT than TiO_2/P3HT. By the same token, in another ternary blend of P3HT:PCBM:GQDs, GQDs reduces the hurdle between the HOMO and LUMO energy levels of P3HT and PCBM brings it to an efficiency of 5.24% [153].

Crystalline silicon (c-Si)/GQDs heterojunction is the next one in organic/inorganic hybrid BHJs. There are a number of surface hanging groups in Si like H-, SiO_x- and CH_3- which leads to pronounced recombination. It is observed that the generation of a heterojunction with GQDs lowers its recombination rate. This is due to the facts such as (i) high gap between the c-Si and GQDs, and (ii) low LUMO level of GQDs as compared to the conduction band (CB) of Si [154].

As propounded above in OSCs, a ternary unit added to an active layer of bulk heterojunctions boosts the PCE of corresponding solar cells. The third unit must hold an ability to enhance light absorption and create interdiffused configuration. Integration of GQDs to a polymer:fullerene based BHJ solar cell uncovers the curtain before the role of GQDs in light absorbing layers. In this ternary unit, GQDs get distributed over fullerene domain and enhances the FRET from GQDs to polymer:fullerene, in addition to the augmented QY and bathochromic shift in the fluorescence emission spectra. All together contribute to improved power conversion efficiency [155]. Based on the above-mentioned facts, it was noted that the design of apposite GQDs will be a perfect choice for GQDs based organic solar cells.

5.3. Light-Emitting Diodes (LEDs)

Basically, LEDs are simple semiconductor (like p-n junctions) devices that emit light when current passes through them in the forward direction [156]. Electroluminescence is the principle behind LEDs which is antithetical to the working principle of solar cells. When current flows along the semiconductor, excitons are generated and instantly the excited electrons recombine with holes releasing light. Schematic representation of the working of LEDs is shown in Figure 27 [156]. Depending upon the materials used, LEDs will emit color of some specific spectral wavelength. The alloys such as aluminum:gallium:indium phosphide (Al alloys give red, orange, and yellow lights) and indium:gallium nitride (In alloys give green, blue and white lights) are used habitually [157, 158]. Over and above ordinary light sources, LEDs possess numerous benefits such as lower power consumption, longer lifetime, smaller size, and faster switching [159].

One of the engrossing class of LEDs is the organic light-emitting diodes (OLEDs) in which organic compounds occupy the position of electroluminescent materials as the emissive layer of the diode. They can be small organic molecule or long chain polymers, which itself acts as a semiconductor. OLEDs are cost effective and thin with wide viewing angle and high contrast and color range [159]. OLEDs find application in visual displays for electronic devices like televisions, mobile phones, digital cameras and so forth. Nowadays GQDs has been considered as a better light emitting layer in OLEDs, sometimes called as Quantum Dot (QD) LEDs [154].

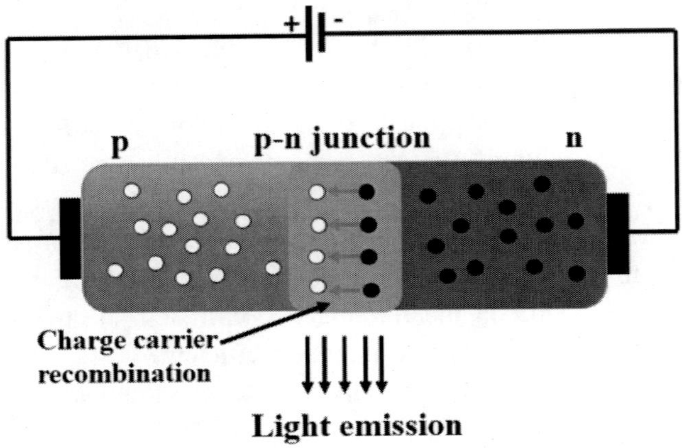

Figure 27. Schematic representation of the working of LEDs.

There is a number of LEDs based on GQDs such as water soluble GQDs [160], octadecylamine-GQDs [161], GQD-agar [162], aniline derivatives functionalized GQDs [163], GQDs cast in UV-curable siloxane [164], oleyl amine capped GQDs [165] and so forth. Water-soluble GQDs without any assistance of additional functionalization or polymers are used for color tunable LEDs [160]. The recombination occurs within GQDs, carries different radiative recombination pathways on distinct excitation energies, which engenders tunable colors that are inevitable for opto-electronic applications. Solvent molecules, ions and functional groups present on the water soluble GQDs influence the defect state emission, while the intrinsic state emission remains unaffected [160]. Coating GQD-agar composite over blue LEDs gives back the white LEDs with a luminous and light conversion efficiency of 42.2 Im W^{-1} and 61% respectively along with 100 hours of color stability [162]. Over and above, GQD based LEDs faced an obstacle of attaining high color purity, which is cured by different aniline derivatives functionalized GQDs [163]. They narrow down the PL linewidths attributed from the origin of new extrinsic energy levels contributed by these aniline derivatives, exhibiting orange, red and green electroluminescence.

Howbeit GQDs-based LEDs exhibit these fascinating properties, yet they are underdeveloped in comparison with inorganic QDs based LEDs. In the context of resource depletion and pollution due to the use of toxic heavy metals, the functionalized GQDs are the promising materials as green sources of light in future visual display units.

Conclusion

In a nutshell, this chapter endeavors an extensive account of the recent advancements in the GQD research with a special focus on bioimaging, biosensing, solar cell, and LED applications. These zero-dimensional materials in the carbon family possess innumerous properties like quantum confinement effect, edge effect, surface states presence and corresponding optical features, biocompatibility and so forth. By the same token, incorporation of other moieties either covalently or non-covalently bound can augment these properties according to our needs. The recent progress on the potential *in vivo* and *in vitro* bioimaging applications of GQDs and their derivatives has been discussed. A correlative and balanced outlook on the various mechanisms involved in the bioimaging applications of GQD based systems has been given, which may facilitate conquering the currently existing hurdles and expedite the fabrication of next-generation GQD derived contrast agents for various biomedical applications. Likewise, various approaches towards the development of GQD based biosensors and fluorescence immunoassays has been addressed. Owing to the superior biocompatibility and water solubility, GQD based biosensors receive much attention and can be utilized for the detection of both small molecules and biological macromolecules.

Moreover, this chapter addresses the successful utilization of GQD based systems in the optoelectronics field due to their unique photoluminescence properties. GQDs behave like p-n junctions, having efficient exciton formation and carrier mobility which leads to better conductivity and enhanced power conversion efficiency of energy devices. In DSSCs, the high surface area, optical path length, good light absorption, and presence of edge functional groups in GQDs favor their use as photoanode, photosensitizer, electrolyte. and counter electrode. Disparate to DSSCs, LEDs rely on reverse working principle, yet GQD based LEDs utilize the same carrier mobility and bandgap tunability. Still GQD based devices are in their childhood, hoping for a better future in the context of other inorganic and organic moieties being toxic and environment pollutants.

References

[1] Tian, P., L. Tang, K. S. Teng, and S. P. Lau. "Graphene quantum dots from chemistry to applications." *Materials today chemistry* 10 (2018): 221-258.

[2] Bak, Sora, Doyoung Kim, and Hyoyoung Lee. "Graphene quantum dots and their possible energy applications: A review." *Current Applied Physics* 16, no. 9 (2016): 1192-1201.

[3] Zhao, Yang, Chuangang Hu, Yue Hu, Huhu Cheng, Gaoquan Shi, and Liangti Qu. "A versatile, ultralight, nitrogen-doped graphene framework." *Angewandte Chemie International Edition* 51, no. 45 (2012): 11371-11375.

[4] Dong, Yongqiang, Jingwei Shao, Congqiang Chen, Hao Li, Ruixue Wang, Yuwu Chi, Xiaomei Lin, and Guonan Chen. "Blue luminescent graphene quantum dots and graphene oxide prepared by tuning the carbonization degree of citric acid." *Carbon* 50, no. 12 (2012): 4738-4743.

[5] Zhao, Changhong, Xuebin Song, Ya Liu, Yifeng Fu, Lilei Ye, Nan Wang, Fan Wang et al. "Synthesis of graphene quantum dots and their applications in drug delivery." *Journal of Nanobiotechnology* 18, no. 1 (2020): 1-32.

[6] Xu, Minghan, Zhaohui Li, Xingzhong Zhu, Nantao Hu, Hao Wei, Zhi Yang, and Yafei Zhang. "Hydrothermal/Solvothermal Synthesis of Graphene Quantum Dots and Their Biological Applications." *Nano Biomedicine & Engineering* 5, no. 2 (2013).

[7] Zhu, Shoujun, Yubin Song, Joy Wang, Hao Wan, Yuan Zhang, Yang Ning, and Bai Yang. "Photoluminescence mechanism in graphene quantum dots: Quantum confinement effect and surface/edge state." *Nano Today* 13 (2017): 10-14.

[8] Hassan, Mahbub, Enamul Haque, Kakarla Raghava Reddy, Andrew I. Minett, Jun Chen, and Vincent G. Gomes. "Edge-enriched graphene quantum dots for enhanced photoluminescence and supercapacitance." *Nanoscale* 6, no. 20 (2014): 11988-11994.

[9] Kalluri, Ankarao, Debika Debnath, Bhushan Dharmadhikari, and Prabir Patra. "Graphene quantum dots: Synthesis and applications." In *Methods in enzymology*, vol. 609, pp. 335-354. Academic Press, 2018.

[10] Gan, Zhixing, Hao Xu, and Yanling Hao. "Mechanism for excitation-dependent photoluminescence from graphene quantum dots and other graphene oxide derivates: consensus, debates and challenges." *Nanoscale* 8, no. 15 (2016): 7794-7807.

[11] Kadian, Sachin, Gaurav Manik, Ashish Kalkal, Manjinder Singh, and Rishi Pal Chauhan. "Effect of sulfur doping on fluorescence and quantum yield of graphene quantum dots: an experimental and theoretical investigation." *Nanotechnology* 30, no. 43 (2019): 435704.

[12] Qian, Zhaosheng, Juanjuan Ma, Xiaoyue Shan, Linxiang Shao, Jin Zhou, Jianrong Chen, and Hui Feng. "Surface functionalization of graphene quantum dots with small organic molecules from photoluminescence modulation to bioimaging applications: an experimental and theoretical investigation." *Rsc Advances* 3, no. 34 (2013): 14571-14579.

[13] Sekiya, Ryo, Yuichiro Uemura, Hideki Murakami, and Takeharu Haino. "White-light-emitting edge-functionalized graphene quantum dots." *Angewandte Chemie* 126, no. 22 (2014): 5725-5729.

[14] Zhang, Ruiqiang, Shifei Qi, Jianfeng Jia, Bryna Torre, Hao Zeng, Haishun Wu, and Xiaohong Xu. "Size and refinement edge-shape effects of graphene quantum dots on UV–visible absorption." *Journal of Alloys and Compounds* 623 (2015): 186-191.

[15] Chakravarty, Chandrima, Poulami Ghosh, Bikash Mandal, and Pranab Sarkar. "Understanding the electronic structure of graphene quantum dot-fullerene nanohybrids for photovoltaic applications." *Zeitschrift für Physikalische Chemie* 230, no. 5-7 (2016): 777-790.

[16] Abbas, Aumber, Tanveer A. Tabish, Steve J. Bull, Tuti Mariana Lim, and Anh N. Phan. "High yield synthesis of graphene quantum dots from biomass waste as a highly selective probe for Fe^{3+} sensing." *Scientific reports* 10, no. 1 (2020): 1-16.

[17] Dorontić, Slađana, Svetlana Jovanović, and Aurelio Bonasera. "Shedding Light on Graphene Quantum Dots: Key Synthetic Strategies, Characterization Tools, and Cutting-Edge Applications." *Materials* 14, no. 20 (2021): 6153.

[18] Soman, Sithara, Aswathy PV and Kala Ramakrishnan. "Covalently modified graphene quantum dot using a thiourea based imprinted polymer for the selective electrochemical sensing of Hg (II) ions." *Journal of Polymer Research* 28, no. 9 (2021): 1-12.

[19] Li, Si-Yu, and Lin He. "Recent progresses of quantum confinement in graphene quantum dots." *Frontiers of Physics* 17, no. 3 (2022): 1-25.

[20] Zhu, Shoujun, Yubin Song, Joy Wang, Hao Wan, Yuan Zhang, Yang Ning, and Bai Yang. "Photoluminescence mechanism in graphene quantum dots: Quantum confinement effect and surface/edge state." *Nano Today* 13 (2017): 10-14.

[21] Jin, Song, Youxuan Ni, Zhimeng Hao, Kai Zhang, Yong Lu, Zhenhua Yan, Yajuan Wei, Ying-Rui Lu, Ting-Shan Chan, and Jun Chen. "A universal graphene quantum dot tethering design strategy to synthesize single-atom catalysts." *Angewandte Chemie International Edition* 59, no. 49 (2020): 21885-21889.

[22] de Mello Donegá, Celso. "Synthesis and properties of colloidal heteronanocrystals." Chemical Society Reviews 40, no. 3 (2011): 1512-1546.

[23] Facure, Murilo HM, Rodrigo Schneider, Luiza A. Mercante, and Daniel S. Correa. "A review on graphene quantum dots and their nanocomposites: from laboratory synthesis towards agricultural and environmental applications." *Environmental Science: Nano* 7, no. 12 (2020): 3710-3734.

[24] Zhu, Shoujun, Yubin Song, Xiaohuan Zhao, Jieren Shao, Junhu Zhang, and Bai Yang. "The photoluminescence mechanism in carbon dots (graphene quantum dots, carbon nanodots, and polymer dots): current state and future perspective." *Nano research* 8, no. 2 (2015): 355-381.

[25] Zhu, Shoujun, Jieren Shao, Yubin Song, Xiaohuan Zhao, Jianglin Du, Lei Wang, Haiyu Wang, Kai Zhang, Junhu Zhang, and Bai Yang. "Investigating the surface state of graphene quantum dots." *Nanoscale* 7, no. 17 (2015): 7927-7933.

[26] Sebastian, Deepa, Sithara Soman and Kala Ramakrishnan. "A nanohybrid system based on covalently functionalized graphene quantum dots with dithienopyrrole

derivative for the sensitive and selective fluorometric detection of Pb^{2+} ions." *Luminescence* 36, no. 7 (2021): 1743-1750.

[27] Sebastian, Deepa, R. Kala, K. P. Neethu Parvathy, and D. P. Savitha. "A fluorescent probe based on visible light-emitting functionalized Graphene Quantum dots for the sensitive and selective detection of Pb (II) ions." *Journal of Materials Science* 56, no. 32 (2021): 18126-18146.

[28] Suzuki, Kaho, Kairi Yamato, Ryo Sekiya, and Takeharu Haino. "Photoluminescence responses of graphene quantum dots toward organic bases and an acid." *Photochemical & Photobiological Sciences* 16, no. 5 (2017): 623-626.

[29] Saleem, Y., L. Najera Baldo, A. Delgado, L. Szulakowska, and P. Hawrylak. "Oscillations of the bandgap with size in armchair and zigzag graphene quantum dots." *Journal of Physics: Condensed Matter* 31, no. 30 (2019): 305503.

[30] Dong, Yongqiang, Congqiang Chen, Xinting Zheng, Lili Gao, Zhiming Cui, Hongbin Yang, Chunxian Guo, Yuwu Chi, and Chang Ming Li. "One-step and high yield simultaneous preparation of single-and multi-layer graphene quantum dots from CX-72 carbon black." *Journal of Materials Chemistry* 22, no. 18 (2012): 8764-8766.

[31] Sudolská, Mária, and Michal Otyepka. "Exact roles of individual chemical forms of nitrogen in the photoluminescent properties of nitrogen-doped carbon dots." *Applied Materials Today* 7 (2017): 190-200.

[32] Dey, Tamal, Subhrajit Mukherjee, Arup Ghorai, Soumen Das, and Samit K. Ray. "Surface state selective tunable emission of graphene quantum dots exhibiting novel thermal quenching characteristics." *Carbon* 140 (2018): 394-403.

[33] Dsouza, Slavia Deeksha, Marius Buerkle, Paul Brunet, Chiranjeevi Maddi, Dilli Babu Padmanaban, Alessio Morelli, Amir Farokh Payam, Paul Maguire, Davide Mariotti, and Vladimir Svrcek. "The importance of surface states in N-doped carbon quantum dots." *Carbon* 183 (2021): 1-11.

[34] Pallikkara, Athira, and Kala Ramakrishnan. "Probing the Photophysics of Covalently and Non-covalently Bonded Graphene Quantum Dots-Tetraaminophenylporphyrin Nanohybrids." *ChemistrySelect* 7, no. 27 (2022): e202201265.

[35] Pallikkara, Athira, Deepa Sebastian, and Kala Ramakrishnan. "An Investigation on the Effect of Extended Conjugation on the Photophysical Properties of Graphene Quantum Dot-Porphyrin Dyads." *ChemistrySelect* 6, no. 43 (2021): 12224-12232.

[36] Kim, Sung, Sung Won Hwang, Min-Kook Kim, Dong Yeol Shin, Dong Hee Shin, Chang Oh Kim, Seung Bum Yang et al. "Anomalous behaviors of visible luminescence from graphene quantum dots: interplay between size and shape." *ACS nano* 6, no. 9 (2012): 8203-8208.

[37] Sebastian, Deepa, Athira Pallikkara, Himanshu Bhatt, Hirendra N. Ghosh, and Kala Ramakrishnan. "Unravelling the Surface-State Assisted Ultrafast Charge Transfer Dynamics of Graphene Quantum Dot-Based Nanohybrids via Transient Absorption Spectroscopy." *The Journal of Physical Chemistry C* 126, no. 27 (2022): 11182-11192.

[38] Cho, Han-Hee, Hyunseung Yang, Dong Jin Kang, and Bumjoon J. Kim. "Surface engineering of graphene quantum dots and their applications as efficient surfactants." *ACS Applied Materials & Interfaces* 7, no. 16 (2015): 8615-8621.

[39] Bao, Lei, Zhi-Ling Zhang, Zhi-Quan Tian, Li Zhang, Cui Liu, Yi Lin, Baoping Qi, and Dai-Wen Pang. "Electrochemical tuning of luminescent carbon nanodots: from preparation to luminescence mechanism." *Advanced materials* 23, no. 48 (2011): 5801-5806.

[40] Luk, C. M., B. L. Chen, K. S. Teng, L. B. Tang, and Shu Ping Lau. "Optically and electrically tunable graphene quantum dot–polyaniline composite films." *Journal of Materials Chemistry C* 2, no. 23 (2014): 4526-4532.

[41] Sun, Hanjun, Nan Gao, Li Wu, Jinsong Ren, Weili Wei, and Xiaogang Qu. "Highly photoluminescent amino-functionalized graphene quantum dots used for sensing copper ions." *Chemistry–A European Journal* 19, no. 40 (2013): 13362-13368.

[42] Henna, T. K., and K. Pramod. "Biocompatibility of graphene quantum dots and related materials." In *Handbook of Biomaterials Biocompatibility*, pp. 353-367. Woodhead Publishing, 2020.

[43] Chong, Yu, Yufei Ma, He Shen, Xiaolong Tu, Xuan Zhou, Jiaying Xu, Jianwu Dai, Saijun Fan, and Zhijun Zhang. "The in vitro and in vivo toxicity of graphene quantum dots." *Biomaterials* 35, no. 19 (2014): 5041-5048.

[44] Shen, Jianhua, Yihua Zhu, Xiaoling Yang, and Chunzhong Li. "Graphene quantum dots: emergent nanolights for bioimaging, sensors, catalysis and photovoltaic devices." *Chemical communications* 48, no. 31 (2012): 3686-3699.

[45] Zhang, Mo, Linling Bai, Weihu Shang, Wenjing Xie, Hong Ma, Yingyi Fu, Decai Fang et al. "Facile synthesis of water-soluble, highly fluorescent graphene quantum dots as a robust biological label for stem cells." *Journal of materials chemistry* 22, no. 15 (2012): 7461-7467.

[46] Kim, Beomsue, Hongmin Kim, Songhui Kim, and Young-ran Hwang. "A brief review of non-invasive brain imaging technologies and the near-infrared optical bioimaging." *Applied Microscopy* 51, no. 1 (2021): 1-10.

[47] Guo, Zhiqian, Sookil Park, Juyoung Yoon, and Injae Shin. "Recent progress in the development of near-infrared fluorescent probes for bioimaging applications." *Chemical Society Reviews* 43, no. 1 (2014).

[48] Pierce, Mark C., David J. Javier, and Rebecca Richards-Kortum. "Optical contrast agents and imaging systems for detection and diagnosis of cancer." *International journal of cancer* 123, no. 9 (2008): 1979-1990.

[49] Jeong, Keunsoo, Youngsun Kim, Chi Soo Kang, Hong-Jun Cho, Yong-Deok Lee, Ick Chan Kwon, and Sehoon Kim. "Nanoprobes for optical bioimaging." *Optical Materials Express* 6, no. 4 (2016): 1262-1279. 16-29.

[50] Rao, Jianghong, Anca Dragulescu-Andrasi, and Hequan Yao. "Fluorescence imaging in vivo: recent advances." *Current opinion in biotechnology* 18, no. 1 (2007): 17-25.

[51] Nagano, Tetsuo. "Development of fluorescent probes for bioimaging applications." *Proceedings of the Japan Academy, Series B* 86, no. 8 (2010): 837-847.

[52] Li, Chunyan, Guangcun Chen, Yejun Zhang, Feng Wu, and Qiangbin Wang. "Advanced fluorescence imaging technology in the near-infrared-II window for

biomedical applications." *Journal of the American Chemical Society* 142, no. 35 (2020): 14789-14804.

[53] Tsuji, Kouichi, Tsuyoshi Matsuno, Yuki Takimoto, Masaki Yamanashi, Noritsugu Kometani, Yuji C. Sasaki, Takeshi Hasegawa et al. "New developments of X-ray fluorescence imaging techniques in laboratory." *Spectrochimica Acta Part B: Atomic Spectroscopy* 113 (2015): 43-53.

[54] Gomez, I. Jennifer, Manuel Vázquez Sulleiro, Anna Doleckova, Nadezda Pizurova, Jirina Medalova, Rajarshi Roy, David Necas, and Lenka Zajickova. "Exploring the emission pathways in nitrogen-doped graphene quantum dots for bioimaging." *The Journal of Physical Chemistry C* 125, no. 38 (2021): 21044-21054.

[55] Kortel, Merve, Bhargav D. Mansuriya, Nicole Vargas Santana, and Zeynep Altintas. "Graphene quantum dots as flourishing nanomaterials for bio-imaging, therapy development, and micro-supercapacitors." *Micromachines* 11, no. 9 (2020): 866.

[56] Lu, Huiting, Wenjun Li, Haifeng Dong, and Menglian Wei. "Graphene quantum dots for optical bioimaging." *Small* 15, no. 36 (2019): 1902136.

[57] Das, Ananda, Kyuyoung Bae, and Wounjhang Park. "Enhancement of upconversion luminescence using photonic nanostructures." *Nanophotonics* 9, no. 6 (2020): 1359-1371.

[58] Sun, Cheng, Florian Figge, John A. McGuire, Qiqi Li, and Liang-shi Li. "Biexciton auger recombination in colloidal graphene quantum dots." *Physical Review Letters* 113, no. 10 (2014): 107401.

[59] Yoon, Hyewon, Minsu Park, Jungmo Kim, Travis G. Novak, Sukki Lee, and Seokwoo Jeon. "Toward highly efficient luminescence in graphene quantum dots for optoelectronic applications." *Chemical Physics Reviews* 2, no. 3 (2021): 031303.

[60] Zhuo, Shujuan, Mingwang Shao, and Shuit-Tong Lee. "Upconversion and downconversion fluorescent graphene quantum dots: ultrasonic preparation and photocatalysis." *ACS nano* 6, no. 2 (2012): 1059-1064.

[61] Li, Xin, Zhuoqi Li, Wupeng Gan, Tongzhou Wang, Songmin Zhao, Ying Lu, Jing Cheng, and Guoliang Huang. "Sensitive and high-resolution subcutaneous fluorescence in vivo imaging using upconversion nanoparticles and microarrays." *Analyst* 138, no. 13 (2013): 3711-3718.

[62] Ji, Yanan, Wen Xu, Nan Ding, Haitao Yang, Hongwei Song, Qingyun Liu, Hans Ågren, Jerker Widengren, and Haichun Liu. "Huge upconversion luminescence enhancement by a cascade optical field modulation strategy facilitating selective multispectral narrow-band near-infrared photodetection." *Light: Science & Applications* 9, no. 1 (2020): 1-12.

[63] Zhang, Youming, Yafei Wang, Jun Song, Junle Qu, Buhong Li, Weiguo Zhu, and Wai-Yeung Wong. "Near-infrared emitting materials via harvesting triplet excitons: molecular design, properties, and application in organic light emitting diodes." *Advanced Optical Materials* 6, no. 18 (2018): 1800466.

[64] Tang, Siyuan, Da Chen, Yongsheng Yang, Changxing Wang, Xiameng Li, Yiru Wang, Chenjie Gu, and Zhen Cao. "Mechanisms behind multicolor tunable Near-

Infrared triple emission in graphene quantum dots and ratio fluorescent probe for water detection." *Journal of Colloid and Interface Science* 617 (2022): 182-192.
[65] Feng, Li-Li, Yong-Xiang Wu, Dai-Liang Zhang, Xiao-Xiao Hu, Jing Zhang, Peng Wang, Zhi-Ling Song, Xiao-Bing Zhang, and Weihong Tan. "Near infrared graphene quantum dots-based two-photon nanoprobe for direct bioimaging of endogenous ascorbic acid in living cells." *Analytical chemistry* 89, no. 7 (2017): 4077-4084.
[66] Kumawat, Mukesh Kumar, Mukeshchand Thakur, Raju B. Gurung, and Rohit Srivastava. "Graphene quantum dots from mangifera indica: application in near-infrared bioimaging and intracellular nanothermometry." *ACS Sustainable Chemistry & Engineering* 5, no. 2 (2017): 1382-1391.
[67] Tang, Libin, Rongbin Ji, Xueming Li, Gongxun Bai, Chao Ping Liu, Jianhua Hao, Jingyu Lin et al. "Deep ultraviolet to near-infrared emission and photoresponse in layered N-doped graphene quantum dots." *ACS nano* 8, no. 6 (2014): 6312-6320.
[68] Gomez, I. Jennifer, Manuel Vázquez Sulleiro, Anna Doleckova, Nadezda Pizurova, Jirina Medalova, Rajarshi Roy, David Necas, and Lenka Zajickova. "Exploring the emission pathways in nitrogen-doped graphene quantum dots for bioimaging." *The Journal of Physical Chemistry C* 125, no. 38 (2021): 21044-21054.
[69] Lee, Bong Han, Ryan Lee McKinney, Md Tanvir Hasan, and Anton V. Naumov. "Graphene quantum dots as intracellular imaging-based temperature sensors." *Materials* 14, no. 3 (2021): 616.
[70] Gao, Tian, Xi Wang, Li-Yun Yang, Huan He, Xiao-Xu Ba, Jie Zhao, Feng-Lei Jiang, and Yi Liu. "Red, yellow, and blue luminescence by graphene quantum dots: syntheses, mechanism, and cellular imaging." *ACS applied materials & interfaces* 9, no. 29 (2017): 24846-24856.
[71] Li Nan, Aung Than, Chencheng Sun, Jingqi Tian, Jie Chen, Kanyi Pu, Xiaochen Dong, and Peng Chen. "Monitoring dynamic cellular redox homeostasis using fluorescence-switchable graphene quantum dots." *ACS nano* 10, no. 12 (2016): 11475-11482.
[72] Karunakaran, Chandran, Raju Rajkumar, and Kalpana Bhargava. "Introduction to biosensors." In *Biosensors and bioelectronics*, pp. 1-68. Elsevier, 2015.
[73] Kaur, Hardeep, Aakash Bhosale, and Suraj Shrivastav. "Biosensors: classification, fundamental characterization and new trends: a review." *Int J Health Sci Res* 8, no. 6 (2018): 315-333.
[74] Suvarnaphaet, Phitsini, and Suejit Pechprasarn. "Graphene-based materials for biosensors: A review." *Sensors* 17, no. 10 (2017): 2161.
[75] Hai, Xin, Ji Feng, Xuwei Chen, and Jianhua Wang. "Tuning the optical properties of graphene quantum dots for biosensing and bioimaging." *Journal of Materials Chemistry B* 6, no. 20 (2018): 3219-3234.
[76] Mobed, Ahmad, Mahdiye Shirafkan, and Saeid Charsouei. "Biosensors modern technology in determination of anti-epileptic drugs (AEDs)." *Clinica Chimica Acta* (2022).
[77] Li, Nana, Ruiyi Li, Zaijun Li, Yongqiang Yang, Guangli Wang, and Zhiguo Gu. "Pentaethylenehexamine and histidine-functionalized graphene quantum dots for ultrasensitive fluorescence detection of microRNA with target and molecular

beacon double cycle amplification strategy." *Sensors and Actuators B: Chemical* 283 (2019): 666-676.

[78] Laurenti, Marco, Miguel Paez-Perez, Manuel Algarra, Paulino Alonso-Cristobal, Enrique Lopez-Cabarcos, Diego Mendez-Gonzalez, and Jorge Rubio-Retama. "Enhancement of the upconversion emission by visible-to-near-infrared fluorescent graphene quantum dots for miRNA detection." *ACS applied materials & interfaces* 8, no. 20 (2016): 12644-12651.

[79] Cunningham, Brian T., and Lance G. Laing. "Advantages and application of label-free detection assays in drug screening." *Expert Opinion on Drug Discovery* 3, no. 8 (2008): 891-901.

[80] Wang, Guang-Li, Xin Fang, Xiu-Ming Wu, Xue-Lian Hu, and Zai-Jun Li. "Label-free and ratiometric detection of nuclei acids based on graphene quantum dots utilizing cascade amplification by nicking endonuclease and catalytic G-quadruplex DNAzyme." *Biosensors and Bioelectronics* 81 (2016): 214-220.

[81] Qian, Xiang, and Ricardo V. Lloyd. "Recent developments in signal amplification methods for in situ hybridization." *Diagnostic Molecular Pathology* 12, no. 1 (2003): 1-13.

[82] Li, Wang, Yong Li, Zhuoliang Liu, Bin Lin, Haibo Yi, Feng Xu, Zhou Nie, and Shouzhuo Yao. "Insight into G-quadruplex-hemin DNAzyme/RNAzyme: adjacent adenine as the intramolecular species for remarkable enhancement of enzymatic activity." *Nucleic acids research* 44, no. 15 (2016): 7373-7384.

[83] Campbell, Elizabeth, Md Tanvir Hasan, Roberto Gonzalez Rodriguez, Giridhar R. Akkaraju, and Anton V. Naumov. "Doped graphene quantum dots for intracellular multicolor imaging and cancer detection." *ACS Biomaterials Science & Engineering* 5, no. 9 (2019): 4671-4682.

[84] Kolesnikova, Sofia, and Edward A. Curtis. "Structure and function of multimeric G-quadruplexes." *Molecules* 24, no. 17 (2019): 3074.

[85] Wei, Yuyun, Di Zhang, Yuxin Fang, Hong Wang, Yangyang Liu, Zhifang Xu, Shenjun Wang, and Yi Guo. "Detection of ascorbic acid using green synthesized carbon quantum dots." *Journal of Sensors* 2019.

[86] Lü, Jian-Ming, Sergiy V. Rosokha, Ivan S. Neretin, and Jay K. Kochi. "Quinones as electron acceptors. X-ray structures, spectral (EPR, UV− vis) characteristics and electron-transfer reactivities of their reduced anion radicals as separated vs contact ion pairs." *Journal of the American Chemical Society* 128, no. 51 (2006): 16708-16719.

[87] Wang, Lirong, Qing Wan, Rongyuan Zhang, Bo Situ, Kaiyuan Ni, Jinhao Gao, Xing Feng et al. "Synergistic Enhancement of Fluorescence and Magnetic Resonance Signals Assisted by Albumin Aggregate for Dual-Modal Imaging." *ACS nano* 15, no. 6 (2021): 9924-9934.

[88] Liu, Hua, Weidan Na, Ziping Liu, Xueqian Chen, and Xingguang Su. "A novel turn-on fluorescent strategy for sensing ascorbic acid using graphene quantum dots as fluorescent probe." *Biosensors and Bioelectronics* 92 (2017): 229-233.

[89] Yang, Lu, Wenfang Deng, Chang Cheng, Yueming Tan, Qingji Xie, and Shouzhuo Yao. "Fluorescent immunoassay for the detection of pathogenic bacteria at the single-cell level using carbon dots-encapsulated breakable organosilica

nanocapsule as labels." *ACS applied materials & interfaces* 10, no. 4 (2018): 3441-3448.

[90] Ekins, Roger, Frederick Chu, and Elizabeth Biggart. "Development of microspot multi-analyte ratiometric immunoassay using dual fluorescent-labelled antibodies." *Analytica Chimica Acta* 227 (1989): 73-96.

[91] Cui, Fangchao, Jian Ji, Jiadi Sun, Jun Wang, Haiming Wang, Yinzhi Zhang, Hong Ding, Yong Lu, Dan Xu, and Xiulan Sun. "A novel magnetic fluorescent biosensor based on graphene quantum dots for rapid, efficient, and sensitive separation and detection of circulating tumor cells." *Analytical and bioanalytical chemistry* 411, no. 5 (2019): 985-995.

[92] Beloglazova, N. V., E. S. Speranskaya, A. Wu, Z. Wang, Melanie Sanders, V. V. Goftman, D. Zhang, I. Yu Goryacheva, and Sarah De Saeger. "Novel multiplex fluorescent immunoassays based on quantum dot nanolabels for mycotoxins determination." *Biosensors and Bioelectronics* 62 (2014): 59-65.

[93] Zou, Fengming, Hongjian Zhou, Tran Van Tan, Jeonghyo Kim, Kwangnak Koh, and Jaebeom Lee. "Dual-mode SERS-fluorescence immunoassay using graphene quantum dot labeling on one-dimensional aligned magnetoplasmonic nanoparticles." *ACS applied materials & interfaces* 7, no. 22 (2015): 12168-12175.

[94] Sharma, Bhavya, Renee R. Frontiera, Anne-Isabelle Henry, Emilie Ringe, and Richard P. Van Duyne. "SERS: Materials, applications, and the future." *Materials today* 15, no. 1-2 (2012): 16-25.

[95] Wolk, Andreas, Marta Rosenthal, Stephan Neuhaus, Klaus Huber, Katharina Brassat, Jörg KN Lindner, Richard Grothe, Guido Grundmeier, Wolfgang Bremser, and René Wilhelm. "A novel lubricant based on covalent functionalized graphene oxide quantum dots." *Scientific reports* 8, no. 1 (2018): 1-9.

[96] Sekar, Rajesh Babu, and Ammasi Periasamy. "Fluorescence resonance energy transfer (FRET) microscopy imaging of live cell protein localizations." *The Journal of cell biology* 160, no. 5 (2003): 629.

[97] Kong, Liyan, Yi Li, Chao Ma, Baihui Liu, and Liang Tan. "Sensitive immunoassay of von Willebrand factor based on fluorescence resonance energy transfer between graphene quantum dots and Ag@ Au nanoparticles." *Colloids and Surfaces B: Biointerfaces* 165 (2018): 286-292.

[98] Li, Nan, Linhong Shi, Xue Zou, Tengfei Wang, Dongmei Wang, Zhengjun Gong, and Meikun Fan. "Fluorescence immunoassay rapid detection of 2019-nCoV antibody based on the fluorescence resonance energy transfer between graphene quantum dots and Ag@ Au nanoparticle." *Microchemical Journal* 173 (2022): 107046.

[99] Garg, Mayank, Reetu Rani, Amit L. Sharma, and Suman Singh. "White graphene quantum dots as electrochemical sensing platform for ferritin." *Faraday Discussions* 227 (2021): 204-212.

[100] Mansuriya, Bhargav D., and Zeynep Altintas. "Applications of graphene quantum dots in biomedical sensors." *Sensors* 20, no. 4 (2020): 1072.

[101] Tang, Libin, Rongbin Ji, Xiangke Cao, Jingyu Lin, Hongxing Jiang, Xueming Li, Kar Seng Teng et al. "Deep ultraviolet photoluminescence of water-soluble self-passivated graphene quantum dots." *ACS nano* 6, no. 6 (2012): 5102-5110.

[102] Liu, Wen-Wen, Ya-Qiang Feng, Xing-Bin Yan, Jiang-Tao Chen, and Qun-Ji Xue. "Superior micro-supercapacitors based on graphene quantum dots." *Advanced Functional Materials* 23, no. 33 (2013): 4111-4122.

[103] Li, Yan, Yang Zhao, Huhu Cheng, Yue Hu, Gaoquan Shi, Liming Dai, and Liangti Qu. "Nitrogen-doped graphene quantum dots with oxygen-rich functional groups." *Journal of the American Chemical Society* 134, no. 1 (2012): 15-18.

[104] Zhao, Xuewei, Yizeng Wu, Yunsong Wang, Huaisheng Wu, Yawei Yang, Zhipeng Wang, Linxiu Dai, Yuanyuan Shang, and Anyuan Cao. "High-performance Li-ion batteries based on graphene quantum dot wrapped carbon nanotube hybrid anodes." *Nano Research* 13, no. 4 (2020): 1044-1052.

[105] Pallikkara, Athira, and Kala Ramakrishnan. "Efficient charge collection of photoanodes and light absorption of photosensitizers: A review." *International Journal of Energy Research* 45, no. 2 (2021): 1425-1448.

[106] https://www.acs.org/content/acs/en/education/resources/highschool/chemmatters/past-issues/archive-2013-2014/how-a-solar-cell-works.html#:~:text=When%20sunlight%20strikes%20a%20solar,to%20the%20p%2Dtype%20layer.

[107] Li, Miaomiao, Wang Ni, Bin Kan, Xiangjian Wan, Long Zhang, Qian Zhang, Guankui Long, Yi Zuo, and Yongsheng Chen. "Graphene quantum dots as the hole transport layer material for high-performance organic solar cells." Physical Chemistry Chemical Physics 15, no. 43 (2013): 18973-18978.

[108] Wu, Wentian, Haixia Wu, Min Zhong, and Shouwu Guo. "Dual role of graphene quantum dots in active layer of inverted bulk heterojunction organic photovoltaic devices." *ACS omega* 4, no. 14 (2019): 16159-16165.

[109] Ameri, Tayebeh, Gilles Dennler, Christoph Lungenschmied, and Christoph J. Brabec. "Organic tandem solar cells: A review." Energy & Environmental Science 2, no. 4 (2009): 347-363.

[110] Li, Yuanchao, Wenyuan Zhang, Xin Li, and Yanling Xu. "Boosting the photoelectric conversion efficiency of DSSCs through graphene quantum dots: insights from theoretical study." *Materials Chemistry Frontiers* 5, no. 15 (2021): 5814-5825.

[111] Mahalingam, Savisha, Abreeza Manap, Azimah Omar, Foo Wah Low, N. F. Afandi, Chin Hua Chia, and Nasrudin Abd Rahim. "Functionalized graphene quantum dots for dye-sensitized solar cell: Key challenges, recent developments and future prospects." Renewable and Sustainable Energy Reviews 144 (2021): 110999.

[112] Chang, Ho, Mu-Jung Kao, Kuohsiu-David Huang, Tung-Jung Hsieh, and Shu-Hua Chien. "Application of TiO_2 nanoparticles coated multi-wall carbon nanotube to dye-sensitized solar cells." *Journal of Nanoscience and Nanotechnology* 10, no. 11 (2010): 7671-7675.

[113] Dembele, Kadiatou Therese, Gurpreet Singh Selopal, Caterina Soldano, Riad Nechache, Julio Cesar Rimada, Isabella Concina, Giorgio Sberveglieri, Federico Rosei, and Alberto Vomiero. "Hybrid carbon nanotubes–TiO_2 photoanodes for high efficiency dye-sensitized solar cells." *The Journal of Physical Chemistry C* 117, no. 28 (2013): 14510-14517.

[114] Salam, Zaahir, E. Vijayakumar, A. Subramania, N. Sivasankar, and Sudhanshu Mallick. "Graphene quantum dots decorated electrospun TiO_2 nanofibers as an effective photoanode for dye sensitized solar cells." *Solar Energy Materials and Solar Cells* 143 (2015): 250-259.

[115] Kumar, D. Kishore, Damaris Suazo-Davila, Desiree García-Torres, Nathan P. Cook, Aruna Ivaturi, Min-Hung Hsu, Angel A. Martí et al. "Low-temperature titania-graphene quantum dots paste for flexible dye-sensitised solar cell applications." *Electrochimica Acta* 305 (2019): 278-284.

[116] Williams, Kenrick J., Cory A. Nelson, Xin Yan, Liang-Shi Li, and Xiaoyang Zhu. "Hot electron injection from graphene quantum dots to TiO_2." *ACS nano* 7, no. 2 (2013): 1388-1394.

[117] Kim, Jinmo, Bongsoo Lee, Young Jun Kim, and Sung Won Hwang. "Enhancement of dye-sensitized solar cells efficiency using graphene quantum dots as photoanode." *Bulletin of the Korean Chemical Society* 40, no. 1 (2019): 56-61.

[118] Mustafa, Muhammad Norhaffis, and Yusran Sulaiman. "Optimization of titanium dioxide decorated by graphene quantum dot as a light scatterer for enhanced dye-sensitized solar cell performance." *Journal of Electroanalytical Chemistry* 876 (2020): 114516.

[119] Fang, Xiaoli, Meiya Li, Kaimo Guo, Jun Li, Muchen Pan, Lihua Bai, Mengdai Luoshan, and Xingzhong Zhao. "Graphene quantum dots optimization of dye-sensitized solar cells." *Electrochimica Acta* 137 (2014): 634-638.

[120] Maity, Nabasmita, Atanu Kuila, Sandip Das, Debasish Mandal, Arnab Shit, and Arun K. Nandi. "Optoelectronic and photovoltaic properties of graphene quantum dot–polyaniline nanostructures." *Journal of Materials Chemistry A* 3, no. 41 (2015): 20736-20748.

[121] Ramakrishnan, Kala and Athira Pallikkara. 2022. "Porphyrin-based photosensitizers for Dye sensitized solar cell applications" In *Photosensitizers and their applications,* edited by Davor Margetic and Renjith Thomas, 193-210, Nova Science Publishers, Inc.

[122] Mihalache, Iuliana, Antonio Radoi, Mihai Mihaila, Cornel Munteanu, Alexandru Marin, Mihai Danila, Mihaela Kusko, and Cristian Kusko. "Charge and energy transfer interplay in hybrid sensitized solar cells mediated by graphene quantum dots." *Electrochimica Acta* 153 (2015): 306-315.

[123] Cole, Jacqueline M., Giulio Pepe, Othman K. Al Bahri, and Christopher B. Cooper. "Cosensitization in dye-sensitized solar cells." *Chemical reviews* 119, no. 12 (2019): 7279-7327.

[124] Subramanian, Alagesan, Zhenghui Pan, Genlan Rong, Hongfei Li, Lisha Zhou, Wanfei Li, Yongcai Qiu et al. "Graphene quantum dot antennas for high efficiency Förster resonance energy transfer based dye-sensitized solar cells." *Journal of Power Sources* 343 (2017): 39-46.

[125] Zamiri, Golnoush, and Samira Bagheri. "Fabrication of green dye-sensitized solar cell based on ZnO nanoparticles as a photoanode and graphene quantum dots as a photo-sensitizer." *Journal of colloid and interface science* 511 (2018): 318-324.

[126] Kumar, M. Praveen, Chisato Takahashi, Subrata Kundu, Tharangattu N. Narayanan, and Deepak K. Pattanayak. "Boron-doped graphene quantum dots: An

efficient photoanode for a dye sensitized solar cell." *New Journal of Chemistry* 43, no. 36 (2019): 14313-14319.

[127] Lee, Eunwoo, Jaehoon Ryu, and Jyongsik Jang. "Fabrication of graphene quantum dots via size-selective precipitation and their application in upconversion-based DSSCs." *Chemical communications* 49, no. 85 (2013): 9995-9997.

[128] Saedi, Afsoon, Ali Mashinchian Moradi, Salimeh Kimiagar, and Homayon Ahmad Panahi. "Efficiency enhancement of dye-sensitized solar cells based on Gracilaria/Ulva using graphene quantum dot." *International Journal of Environmental Research* 14, no. 4 (2020): 393-402.

[129] Li, Dongmei, Da Qin, Minghui Deng, Yanhong Luo, and Qingbo Meng. "Optimization the solid-state electrolytes for dye-sensitized solar cells." *Energy & environmental science* 2, no. 3 (2009): 283-291.

[130] Porfarzollah, Ali, Rahim Mohammad-Rezaei, and Massoumeh Bagheri. "Ionic liquid-functionalized graphene quantum dots as an efficient quasi-solid-state electrolyte for dye-sensitized solar cells." *Journal of Materials Science: Materials in Electronics* 31, no. 3 (2020): 2288-2297.

[131] Wang, Yudi, Chunyu Zhao, Da Qin, Mingxing Wu, Wei Liu, and Tingli Ma. "Transparent flexible Pt counter electrodes for high performance dye-sensitized solar cells." Journal of Materials Chemistry 22, no. 41 (2012): 22155-22159.

[132] Wang, Chun-Chieh, and Shih-Yuan Lu. "Carbon black-derived graphene quantum dots composited with carbon aerogel as a highly efficient and stable reduction catalyst for the iodide/tri-iodide couple." *Nanoscale* 7, no. 3 (2015): 1209-1215.

[133] Chang, Quanhong, Zhoujing Ma, Jinzhong Wang, Panpan Li, Yuan Yan, Wangzhou Shi, Qi Chen, Yuewu Huang, and Lei Huang. "Hybrid Graphene Quantum Dots@ Graphene Foam Nanosheets for Dye-Sensitized Solar Cell Electrodes." *Energy Technology* 4, no. 2 (2016): 256-262.

[134] Brownson, Dale AC, and Craig E. Banks. "The electrochemistry of CVD graphene: progress and prospects." *Physical Chemistry Chemical Physics* 14, no. 23 (2012): 8264-8281.

[135] Yu, Chang, Zhiqiang Liu, Yiwen Chen, Xiangtong Meng, Mingyu Li, and Jieshan Qiu. "CoS nanosheets-coupled graphene quantum dots architectures as a binder-free counter electrode for high-performance DSSCs." *Science China Materials* 59, no. 2 (2016): 104-111.

[136] Liu, Tao, Kun Yu, Lina Gao, Hui Chen, Ning Wang, Luhan Hao, Tingxi Li, Hongcai He, and Zhanhu Guo. "A graphene quantum dot decorated SrRuO 3 mesoporous film as an efficient counter electrode for high-performance dye-sensitized solar cells." *Journal of Materials Chemistry A* 5, no. 34 (2017): 17848-17855.

[137] Chen, Lijia, Chun Xian Guo, Qiaoming Zhang, Yanlian Lei, Jiale Xie, Shujing Ee, Guanhong Guai, Qunliang Song, and Chang Ming Li. "Graphene quantum-dot-doped polypyrrole counter electrode for high-performance dye-sensitized solar cells." *ACS applied materials & interfaces* 5, no. 6 (2013): 2047-2052.

[138] Kundu, Sumana, P. Sarojinijeeva, R. Karthick, G. Anantharaj, G. Saritha, Rajesh Bera, S. Anandan et al. "Enhancing the Efficiency of DSSCs by the Modification

of TiO_2 Photoanodes using N, F and S, co-doped Graphene Quantum Dots." *Electrochimica Acta* 242 (2017): 337-343.

[139] Dinari, Mohammad, Mohamad Mohsen Momeni, and Meysam Goudarzirad. "Dye-sensitized solar cells based on nanocomposite of polyaniline/graphene quantum dots." *Journal of materials science* 51, no. 6 (2016): 2964-2971.

[140] Jahantigh, Farhad, SM Bagher Ghorashi, and Amir Bayat. "Hybrid dye sensitized solar cell based on single layer graphene quantum dots." *Dyes and Pigments* 175 (2020): 108118.

[141] Rahman, Mati Ur, Fengyan Xie, Yafeng Li, and Mingdeng Wei. "ZnO nanosheets encapsulating graphene quantum dots with enhanced performance for dye-sensitized solar cell." *Journal of Electroanalytical Chemistry* 840 (2019): 160-164.

[142] https://en.wikipedia.org/wiki/Organic_solar_cell.

[143] Wadsworth, Andrew, Zeinab Hamid, Jan Kosco, Nicola Gasparini, and Iain McCulloch. "The bulk heterojunction in organic photovoltaic, photodetector, and photocatalytic applications." *Advanced Materials* 32, no. 38 (2020): 2001763.

[144] Li, Yan, Yue Hu, Yang Zhao, Gaoquan Shi, Lier Deng, Yanbing Hou, and Liangti Qu. "An electrochemical avenue to green-luminescent graphene quantum dots as potential electron-acceptors for photovoltaics." *Advanced materials* 23, no. 6 (2011): 776-780.

[145] Günes, Serap, Helmut Neugebauer, and Niyazi Serdar Sariciftci. "Conjugated polymer-based organic solar cells." *Chemical reviews* 107, no. 4 (2007): 1324-1338.

[146] Gupta, Vinay, Tanvi Upreti, and Suresh Chand. "Graphene Quantum Dot-Based Organic Solar Cells." *Quantum Dot Solar Cells* (2014): 255-268.

[147] Gupta, Vinay, Neeraj Chaudhary, Ritu Srivastava, Gauri Datt Sharma, Ramil Bhardwaj, and Suresh Chand. "Luminscent graphene quantum dots for organic photovoltaic devices." *Journal of the American Chemical Society* 133, no. 26 (2011): 9960-9963.

[148] Wu, Wentian, Haixia Wu, Min Zhong, and Shouwu Guo. "Dual role of graphene quantum dots in active layer of inverted bulk heterojunction organic photovoltaic devices." *ACS omega* 4, no. 14 (2019): 16159-16165.

[149] Li, Miaomiao, Wang Ni, Bin Kan, Xiangjian Wan, Long Zhang, Qian Zhang, Guankui Long, Yi Zuo, and Yongsheng Chen. "Graphene quantum dots as the hole transport layer material for high-performance organic solar cells." *Physical Chemistry Chemical Physics* 15, no. 43 (2013): 18973-18978.

[150] http://depts.washington.edu/cmditr/modules/opv/organic_heterojunctions_in_solar_cells.html.

[151] Zhang, Jinshui, Mingwen Zhang, Rui-Qing Sun, and Xinchen Wang. "A facile band alignment of polymeric carbon nitride semiconductors to construct isotype heterojunctions." *Angewandte Chemie International Edition* 51, no. 40 (2012): 10145-10149.

[152] Qin, Yuancheng, Yuanyuan Cheng, Longying Jiang, Xiao Jin, Mingjun Li, Xubiao Luo, Guoqing Liao, Taihuei Wei, and Qinghua Li. "Top-down strategy toward versatile graphene quantum dots for organic/inorganic hybrid solar cells." *ACS Sustainable Chemistry & Engineering* 3, no. 4 (2015): 637-644.

[153] Li, Fushan, Lijie Kou, Wei Chen, Chaoxing Wu, and Tailiang Guo. "Enhancing the short-circuit current and power conversion efficiency of polymer solar cells with graphene quantum dots derived from double-walled carbon nanotubes." *NPG Asia Materials* 5, no. 8 (2013): e60-e60.

[154] Gao, Peng, Ke Ding, Yan Wang, Kaiqun Ruan, Senlin Diao, Qing Zhang, Baoquan Sun, and Jiansheng Jie. "Crystalline Si/graphene quantum dots heterojunction solar cells." *The Journal of Physical Chemistry C* 118, no. 10 (2014): 5164-5171.

[155] Sung, Sae Jin, Jae Ho Kim, Se Hoon Gihm, Jisoo Park, Young Shik Cho, Seung Jae Yang, and Chong Rae Park. "Revisiting the Role of Graphene Quantum Dots in Ternary Organic Solar Cells: Insights into the Nanostructure Reconstruction and Effective Förster Resonance Energy Transfer." *ACS Applied Energy Materials* 2, no. 12 (2019): 8826-8835.

[156] https://byjus.com/physics/light-emitting-diode/#:~:text=How%20do%20LEDs%20work%3F,in%20the%20form%20of%20photons.

[157] Phillips, Julia M., Michael E. Coltrin, Mary H. Crawford, Arthur J. Fischer, Michael R. Krames, Regina Mueller-Mach, Gerd O. Mueller et al. "Research challenges to ultra-efficient inorganic solid-state lighting." *Laser & Photonics Reviews* 1, no. 4 (2007): 307-333.

[158] Liu, Zhaojun, Chun-Ho Lin, Byung-Ryool Hyun, Chin-Wei Sher, Zhijian Lv, Bingqing Luo, Fulong Jiang et al. "Micro-light-emitting diodes with quantum dots in display technology." *Light: Science & Applications* 9, no. 1 (2020): 1-23.

[159] https://en.wikipedia.org/wiki/Light-emitting_diode.

[160] Kumar, Gundam Sandeep, Umamahesh Thupakula, Piyush Kanti Sarkar, and Somobrata Acharya. "Easy extraction of water-soluble graphene quantum dots for light emitting diodes." *RSC Advances* 5, no. 35 (2015): 27711-27716.

[161] Kim, Dae Hun, and Tae Whan Kim. "Ultrahigh current efficiency of light-emitting devices based on octadecylamine-graphene quantum dots." *Nano Energy* 32 (2017): 441-447.

[162] Luk, C. M., L. B. Tang, W. F. Zhang, Siu Fung Yu, K. S. Teng, and Shu Ping Lau. "An efficient and stable fluorescent graphene quantum dot–agar composite as a converting material in white light emitting diodes." *Journal of Materials Chemistry* 22, no. 42 (2012): 22378-22381.

[163] Kwon, Woosung, Young-Hoon Kim, Ji-Hee Kim, Taehyung Lee, Sungan Do, Yoonsang Park, Mun Seok Jeong, Tae-Woo Lee, and Shi-Woo Rhee. "High color-purity green, orange, and red light-emitting diodes based on chemically functionalized graphene quantum dots." *Scientific reports* 6, no. 1 (2016): 1-10.

[164] Yin, Luqiao, Jipeng Zhou, Weitao Li, Jianhua Zhang, and Liang Wang. "Yellow fluorescent graphene quantum dots as a phosphor for white tunable light-emitting diodes." *RSC advances* 9, no. 16 (2019): 9301-9307.

[165] Kwon, Woosung, Young-Hoon Kim, Chang-Lyoul Lee, Minkyung Lee, Hee Cheul Choi, Tae-Woo Lee, and Shi-Woo Rhee. "Electroluminescence from graphene quantum dots prepared by amidative cutting of tattered graphite." *Nano letters* 14, no. 3 (2014): 1306-1311.

Chapter 4

Recent Progress in Graphene Quantum Dots for Energy Applications and Storage Devices: A Review

C. O. Edet[1,2,*]
E. P. Agbo[2,†]
H. O. Edet[3,‡]
Nnanake-Abasi O. Offiong[4,§]
John B. Edet[5,‖]
and E. B. Ettah[2,¶]

[1]Faculty of Electronic Engineering Technology, Universiti Malaysia Perlis, Kangar, Perlis, Malaysia
[2]Department of Physics, Cross River University of Technology, Calabar, Nigeria
[3]Computational and Bio-Simulation Research Group, University of Calabar, Calabar, Nigeria
[4]Department of Chemical Sciences, Topfaith University, Mkpatak, Nigeria
[5]Department of Chemical and Process Engineering, University of Strathclyde, Glasgow, United Kingdom

[*] Corresponding Author's Email: collinsokonedet@crutech.edu.ng.
[†] Corresponding Author's Email: emmanuelpaulagbo@gmail.com.
[‡] Corresponding Author's Email: henryokonedet98@gmail.com.
[§] Corresponding Author's Email: no.offiong@topfaith.edu.ng.
[‖] Corresponding Author's Email: john.edet@strath.ac.uk.
[¶] Corresponding Author's Email: emmanuelettah@gmail.com.

In: Graphene Quantum Dots (GQDs)
Editor: Aleksey Kuznetsov
ISBN: 979-8-89113-638-0

Abstract

Graphene quantum dots (GQDs) have their features as a derivative of both graphene and carbon dots (CDs); this gives them unique physical and chemical properties leading to a wide range of applications. In this chapter, we present a review of the utilization of GQDs for energy applications and storage devices. The review addresses recent developments and progress so far as well as the knowledge trends. The different properties of GQDs as related to energy applications are presented. Finally, the prospects and potentials of this development are described.

1. Introduction

Graphene is not only the thinnest but also one of the strongest materials yet created [1]. It is analogous to an infinite 2D monolayer of hexagonal sp^2 bonded carbon network with zero band gap features due to delocalized electrons [2]. As a result of this semi-metallic property, electrons pass through graphene with an almost zero impedance, resulting in better electrical conductivity than copper [3]. Despite graphene having various impressive properties like lightweight, high tensile strength, and thermal conductivity, its synthesis remains the most difficult engineering challenge to overcome before graphene can be used in industry; this is due to the fact that the creation of graphene can only be achieved in very minute amounts; it must then be separated from the other reaction products. There is currently no straightforward manufacturing process that yields significant amounts of high-quality graphene [4]. The graphene industry's major obstacles will be achieving volume manufacturing in the short time [5]. The uniformity of the materials and the cost of production is one of the main concerns. Indeed, creating consistent and homogenous graphene dispersions without agglomeration remains a big difficulty [6]. It has become adamant to investigate the applications of Graphene Quantum Dots (GDQs) due to their recent discovery as a member of the carbon family [7]; the rapid growth in their preparation methods makes studying them very vital for various applications [8, 9].

Carbon, as we know, is one of the most plentiful elements, and has been studied extensively since the 19th century [10–12], with a steady increase in research. These aforementioned works have progressed and yielded to the discovery of GQDs, and although this discovery and application seems promising, there have been a lot of discovered limitations for graphene as earlier mentioned. One of these limitations is its low absorption ability and zero bandgap.

The preparation of GQDs is a subject with depth on its own. Basically, there are two approaches to their preparation – the bottom-up and the top-down synthetic approaches [13, 14]. In a nutshell, top-down synthetic approach entails using highly concentrated acids to breakdown big carbon-based substances like carbon fibres [15], graphene oxide (GO) [16], coal [17, 18], fullerenes [19], graphite [20], etc., into tiny fragments/atomic layers. This approach produces graphene from converting graphite through processes like mechanical depilation [21], liquid-phase exfoliation (LPE) [8], electrochemical depilation methods [22], and chemical oxidation-reduction [23]. Bottom-up synthetic approach has to do with the production of graphene molecularly from low molecular organic (carbon) precursors [24] using chemical vapour decomposition (CVD) or epitaxial grown on a substrate. Graphene created by this method usually have large surface area and high quality, with the method presenting the greatest potential for graphene to be mass-produced. One drawback of this approach is that it has obstacles like low yields as a result of purification challenges in the removal of small organic materials that are unreacted [6]. Readers seeking to get a broader and in-depth knowledge on the synthesis of GQDs can consult quality review papers [25–29].

The purpose of this review arose from the limitation of storage and energy devices that are electrochemically produced. Global warming and other factors that arise as a result of not holistically tackling climate change will always be a major subject area for sustainable development [30]. Studies have shown [31–35] that the climate of a region is directly affected by the long-term meteorological conditions. As an extension, other studies have proven that development of sustainable energy devices will indirectly have an effect on the long-term climate and energy development of a region [36–38]. In fact, the main purpose of the meteorology and climatology is to use the observation results to suggest sustainable technologies and methods to curb or utilise the varying climate as the case may be. That is why it is expedient that efficient energy and storage devices are developed emphasizing the reason why GQDs are being studied as an innovative material for these applications. We have collected results from extensive literature and presented them empirically in the following sections, drawing inferences on their various applications as energy and storage devices.

2. Applications of Graphene Quantum Dots

GQDs are recent innovations which was developed in 2008 due to the need for materials with superior properties to the conventional graphene and quantum dots from inorganic sources [39, 40]. This modification was necessary as graphene was observed to have zero band gap and low absorptivity, while high cost and toxicity were the major concerns of inorganic quantum dots which limits its application (Figure 1). The band gap of these materials is one of the determinants of photoluminescence (PL) properties which is influenced by size, shape, and fraction of sp^2 site. The ability to modify the sp^2 domain of these materials implies that size and shape, which are key determinants of photoluminescence, can be tuned to suit the specific application which is highly desirable in industry. GQDs have an average size range of 2 nm – 50 nm and control over this property is relevant for specific applications [6].

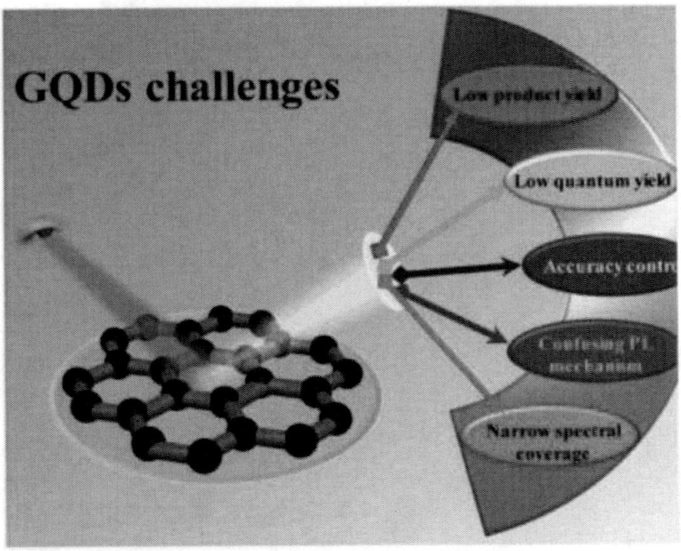

Figure 1. Recent Challenges of GQDs. Reproduced from [40]. Copyright © 2022 Elsevier B.V. https://www.sciencedirect.com/.

Interestingly, this control has been achieved through variation in parameters such as temperature, catalyst, time of growth, source concentration among others during synthesis [40]. Besides, the structural modification that resulted in the fabrication of GQDs involve introduction of graphene lattice with less than 10 layers thick and smaller than 10 nm inside the dots [40]. This resulted in unique properties such as improved solubility, reduced toxicity, surface grafting, inertness and excellent fluorescence compared to other materials such as carbon nanotubes. These unique properties can be further enhanced in the form of doping with heteroatoms or forming composite with other oxides such as tin oxides which have been reported to have good stability and cycling performance resulting in high coulombic efficiency [41]. These characteristics have been exploited in diverse applications such as fabrication of Li-ion batteries, solar cells, photocatalyst, LEDs, capacitors/supercapacitors, and sensors.

2.1. Application of GQDs in Li-Ion Batteries

Lithium-ion battery is one of the energy sources consisting of a graphite anode and a cathode made of a transition metal. This class of batteries have been thoroughly investigated and found to be associated with high energy and power density recharging, durable and possess flexible designs suitable for use in smart electronic storage device and hybrid electric vehicles [42–47]. The performance of Li-ion battery is dependent on size of the nanomaterial as this determines the surface area, diffusion rate and length [44]. Although silicon has been recognized and widely accepted as the anode material with the most promising attributes that fit the requirement for Li-ion battery, the significant variation in volume (up to 259%) during the charge-discharge process culminating in pulverization and rapid capacity loss in cycling has necessitated further search of materials with better characteristics to overcome this limitation [48, 49]. Therefore, this requirement fits into the properties of GQDs which makes them an excellent material of interest for Li-ion batteries.

Computational and experimental investigation into the mechanism of interaction between lithium ions and hexagonal GQDs of different size such as coronene ($C_{24}H_{12}$), circumcoronene ($C_{54}H_{18}$), and circumcircum-coronene ($C_{96}H_{24}$) have been carried out to assess the suitability of different forms of GQDs in Li-ion batteries [50, 51]. These studies indicated that both lithium atom and ion prefer binding to the central portions and hollow site of the hexagonal unit of coronene [52] but prefer adsorbing to the outermost perimeter and hollow sites of circumcoronene [44, 53]. The adsorption of single and multiple lithium ions to GQDs including the loading capacity have been extensively studied and found to have significant influence on their performance as a battery [44]. Also, strong binding of the lithium ions to

GQDs implies increase in adsorption energy, which may likely decrease as more lithium ions adsorb on the material due to ionic repulsion. Figure 2 shows the possibility of incorporating GQDs with the anodes of Li-ion batteries as studied by [54]. However, theoretical calculations based on density functional theory have suggested that the complicated interactions between graphene sheets and multiple lithium ions makes the binding energy independent of the number of adsorbed lithium ion [55].

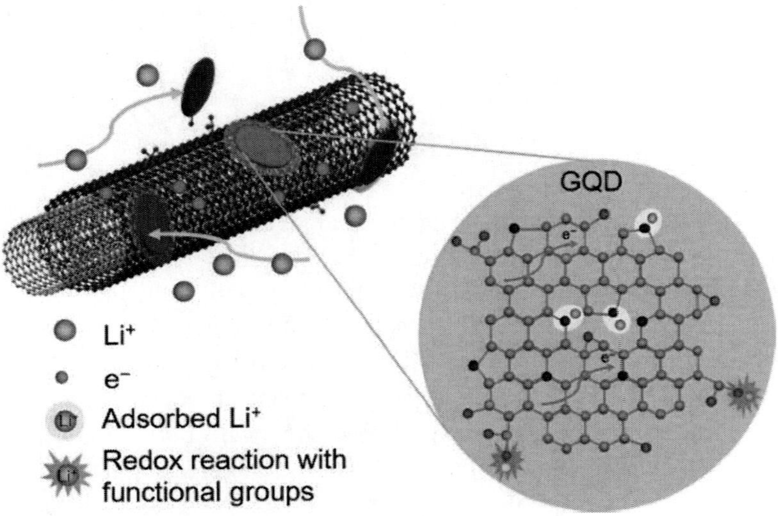

Figure 2. The application of novel carbonaceous Li-ion battery anodes incorporating graphene quantum dots (GQDs). This is made possible through the fabrication of a freestanding all-carbon electrode based on a porous carbon nanotube (CNT) sponge via a facile in-situ hydrothermal deposition technique, creating coaxial structure of GQD-coated CNTs (GQD@CNTs) through electrostatic interaction and π-π stacking with tunable loading and functionalization. Reproduced from [54]. Copyright 2022. Springer Nature, Switzerland AG. Part of Springer Nature. https://www.springernature.com/.

In addition, the doping of sulphur and nitrogen to GQDs have been reported to improve the electron transfer and storage capacity of lithium-ion batteries, even though they do not appear to influence the electrolyte transport [50]. The introduction of these heteroatoms to GQDs is likely to alter the orientation of the carbon framework and adjust charge distribution as well as the spins of the doped sites, resulting in improved performance compared to single GQDs. In another study, the introduction of cobalt sulphide nanoparticles to GQDs was reported to influence the cycle processes and maintains high electrical conductivity of the electrode [56]. In addition, the composite formed from silicon nanoparticles and GQD functionalized with phenylalanine was reported to create a wide energy level of electron and has a well-defined steric structure which enhances electron transfer and electrolyte transport, resulting in improved electrochemical performance [49]. A summary of the modifications to GQD and the resulting effect on their performance is presented in Table 1. Despite the importance of these structural modifications which has evidence of improved properties in Li-ion batteries, it is important to note that in building GQDs or composite form for specific application, the cost associated with these modifications for improved performance should be taken into consideration. This balance between cost and performance is highly desirable in industry and can be achieved by selecting materials that are easily accessible and affordable but have comparable or superior performance as the conventional materials.

Table 1. Properties of modified graphene quantum dots in Li-ion battery

Modifications	Performance		References
	Capacitance ($mAhg^{-1}$)	Coulombic Efficiency (%)	
LTO/S, N-GQD	254.2	96.9, 2000 cycles	[50]
PF-GQD@SiNp	4066	92, 100 cycles	[49]
SiAl/GQDs	2229.5	88.9, 120/150 cycles	[57, 58]
C-GQDs/α-Fe_2O_3	1582.5	98, 5 cycles	[9]
VO_2-GQD	421	99	[59]
SnO2/GQD	797	95	[41]

2.2. Application of GQDs in Solar Cells

The shift in band gap obtained by converting graphene to GQDs resulting in enhanced photoluminescence has been one of the reasons for the use of GQDs in solar cells. Besides, they are good electron acceptors suitable for use in dye-sensitized solar cells [60]. Dye-sensitized solar cells (DSSCs) have the problem of reduced light absorption which limits their power conversion efficiency [9, 61, 62]. Therefore, titanium oxide has been widely reported as the photoanode suitable for meeting this deficiency in solar cells due to its wide band gap [61]. Other authors have reported that the addition of titanium dioxide nanoparticles to single-layer graphene quantum dots has resulted in control over the optical properties and energy states resulting in improved power conversion efficiency for application in DSSC [63]; in extension, research has also shown that carbon-doped zinc-oxide (C-ZnO) has increased the potential for GQD application in solar cells (Figure 3).

Figure 3. GQDs sensitized C-ZnO nanotaper photoanodes for solar cells application. The Photoelectrochemical properties of taper-like zinc-oxide (ZnO) nanorod photoanodes by carbon-doping as well as sensitization with sulphur (S), nitrogen (N) co-doped GQDs. Reproduced from [64]. Copyright 2022 by the American Association for the Advancement of Science (AAAS) https://www.eurekalert.org/.

Apart from the significant role of graphene quantum dots in dye sensitized solar cells, GQDs have also been used in inverted polymer solar cells (IPSCs). This is possible because of the wide band gap of about 3.3 eV, which matches the criteria for IPSCs [62]. It has been reported that the incorporation of GQD to cesium carbonate resulted in enhancement of the power conversion efficiency by 22% when compared to IPSC with just cesium carbonate [62]. In addition, Sharma and Jha [65] reported that doping of GQDs using adatoms such as nitrogen, boron, and phosphorus at the edges where GQD have been functionalized with carboxylic groups have resulted in 22-30% increase in efficiency, with phosphorus being the main contributor since it donates more electrons to the surface of the titanium surface compared to other adatoms. Furthermore, one of the requirements expected of photoanodes for solar cell applications is excellent photoelectrochemical properties. This is defined by high photoconversion efficiency, short circuit current and good incident photon to current conversion efficiency, which were reported to be achieved when zinc oxide nanotapes were combined with nitrogen doped graphene quantum dots [66]. Other features of modified GQDs reported in the literature as well as the resulting characteristics for solar cell applications are presented in Table 2.

Table 2. Properties of modified GQDs for solar cell applications

Modifications	Properties				References
	J_{sc} (mAcm^{-2})	V_{oc} (eV)	PEC (%)	Fill Factor	
TiO_2-GQD	11.54	0.73	4.2-4.43	0.527	[60]
GQD	19.5-22.62	0.65	6.9-7.82	0.53	[61]
N719/TiO2-GQD	20.03	0.73	8.92	0.61	[63]
GQDs–Cs_2CO_3	9.04	0.585	3.17	0.6	[62]
GQDs	4.70	0.509	0.79	0.332	[62]
COOH-GQD, N-GQD, B-GQD and P-GQD	0.474, 0.596, 0.579 and 0.721	1.625, 1.571, 1.632 and 1.39	0.7, 0.86, 0.87 and 0.909	0.918, 0.916, 0.919 and 0.908	[65]

2.3. Application of GQDs as a Photocatalyst

The suitability of GQDs as a photocatalyst for splitting of water into its components has been widely explored due to their high surface area, oxygen-rich surface, negligible toxicity, adjustable bandgap as well as reduced cost compared to other materials [67]. The chemistry of water splitting can either involve a two-electron process resulting in formation of hydrogen and peroxide or a four-electron process yielding hydrogen and oxygen [68]. Both processes require a catalyst with good optical properties to enhance the process. The high surface plasmon resonance property of modified GQDs has been exploited for this purpose. Liang et al. [67] recently reported that nitrogen doped GQDs are capable of splitting water molecule to hydrogen and peroxides, which does not require gas separation usually encountered with conventional catalysts that always yield hydrogen and oxygen products. Figure 4 shows how controlling nitrogen doping of GQDs (N-GQDs) can be applied in electrocatalytic and photo luminescent applications through laser ablation in liquids [69].

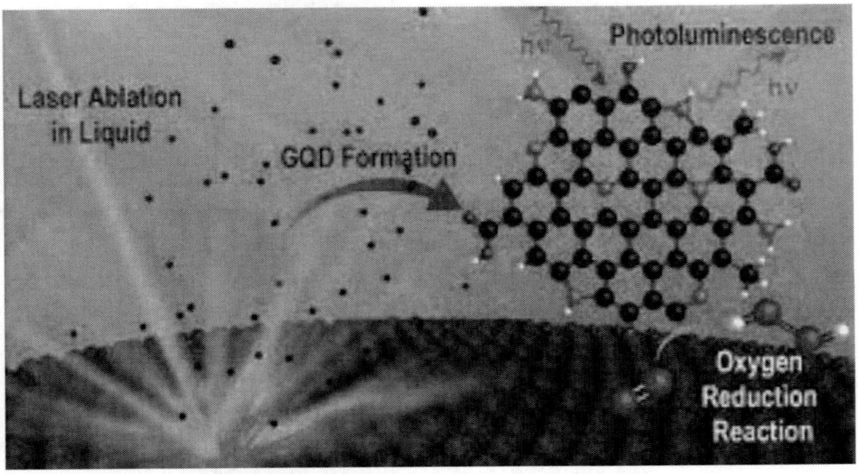

Figure 4. Laser ablation in aqueous solutions (liquids) as a promising method to control nitrogen-doped graphene quantum dots (N-GQDs) to prepare nanomaterials. *Reproduced from* [69]. *Copyright 2022 by ACS Publications https://pubs.acs.org/.*

In addition, modified GQDs have been used as a photocatalyst for the remediation of chemical contaminants from aqueous solution due to their ability to break down toxic chemical compounds to less harmful components. For instance, zero-dimensional GQDs were decorated on carbon nitride nanorods and found to be effective for the removal of antibiotics such as tetracycline and oxytetracycline from water [70]. The authors attributed this activity to the presence of photo induced holes which is promoted by high crystalline level, improved light absorption as well as staggered band alignment. When it comes to remediation of chemical contaminant from aqueous solution, sometimes GQDs may not be efficient when used independently but may enhance the remediation potential of other materials used for the same purpose. For instance, in 2018, it was reported that GQDs were not quite successful in remediating methylene blue from aqueous solution; however, when it was used to surround zinc oxide nanorods (ZnO-NR), it decreased its hole recombination rate with a resulting increase in photocatalytic activity of the composite for the removal of methylene blue from aqueous solution [71, 72]. It is possible to regulate this remediation activity for methylene blue using different polymer modified GQDs [73]. A summary of the importance of modified GQDs in remediating different chemical contaminants from the environment is presented in Table 3.

Table 3. Properties of modified GQD for photocatalytic applications

Modified GQDs	Application	Performance		References
N-GQD@AU-NR	Water splitting	Rate of H_2 evolution: 65.6 mmol g-1 h-1	Rate of H_2O_2 generation: 49.7 mmol g^{-1} h^{-1}	[67]
GQD/g-CNNR	Removal of antibiotics	TOC removal efficiency: 68.1%,	COD removal efficiency: 56.4%	[70]
GQD-Fe-TiO_2	Removal of Rhodamine blue RB5 dye	Removal efficiency: 50 – 97.4%	-	[74]
ZnO-NR/GQD	Removal of methylene blue and Carbendazim	Removal efficiency: 26-95%	-	[75]
S, N-GQD/P25	Water splitting	Rate of H_2 generation: 1.7-5.7 μmol/h	-	[76]

2.4. Capacitors/Supercapacitors

The demand for energy storage and environmental significance has grown worldwide, and sustainable energy supplies have proven to be essential. Although, the application of super capacitors which exhibits the potential for the storage of energy has been a challenge [77]. The discovery of graphene has piqued the interest of experts from a variety of fields. Graphene has been predicted and confirmed as a fantastic contender for opto-electrical applications due to its ultra-high carrier mobility and linear dispersion band structure [78]. Graphene-based materials are unique materials for applications in super capacitors and energy storage devices due to their high conductivity properties and good chemical stability [40]. GQDs are often used as a replacement for activated carbon in super capacitors with relative surface area of 2630 m^2/g [79].

Capacitors and supercapacitors have attracted considerable attentions due to their high charge/discharge rates and long-life cycle performance [80]. GQDs can be made through exfoliation and the resultant graphene becomes hydrophobic [81]. Supercapacitor electrodes made of graphene show a high power density and a fast charge transfer.

A schematic diagram of the various methods for producing carbon-based quantum dots (CBQDs) for use in supercapacitors is shown in Figure 5. A review by Shaker et al. [82] showed how CBQDs can surmount the limits of today's supercapacitors. These materials can contribute to emerging supercapacitors with much higher specific capacitance, energy density, or power density due to the combinatorial effects of enhanced electrochemical reactivity, electrical conductivity, and charge accumulation on the surface.

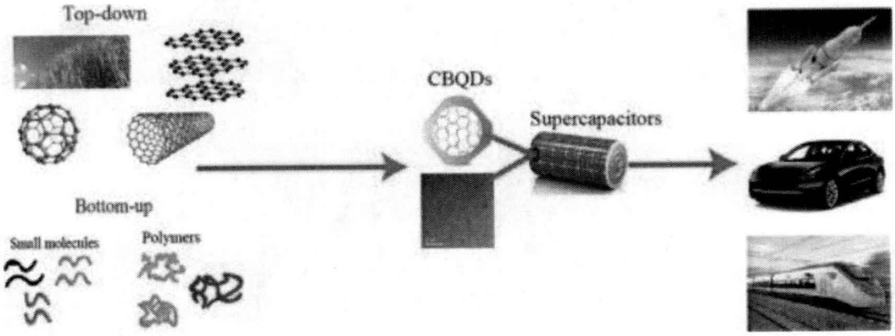

Figure 5. A schematic representation of bottom-up and top-down approaches for making carbon-based quantum dots (CBQDs) and their applications in supercapacitors. *Reproduced from* [82]. *Copyright © 2022 Elsevier B.V. https://www.sciencedirect.com/.*

Recently, [83], a fabrication process was developed for highly porous GO and reduced graphene oxide (rGO) fibers and yarns by taking the advantage of the intrinsic soft assembly behaviour and proposed its use for storage devices. Zhang et al. [84] have shown that GQD electrode material displays an ideal electric double-layer capacitance behaviour such as a high specific capacitance of 296.7 F g^{-1} and a satisfactory density of 41.2 W h kg^{-1} at 1 A g^{-1}. Liu et al. [85] reported that GQDs micro-supercapacitor has a high-rate capability of up to 1000 V s-1, as well as an excellent power response and a short relaxation constant. In aqueous electrolyte, GQDs/MnO_2 asymmetric supercapacitors are also built with GQDs as the positive electrode and GQDs as the negative electrode. The findings suggested that GQDs could be used in micro power sources and micro energy storage devices [86].

2.5. LEDs

Light emitting diodes (LEDs) have a number of advantages over incandescent light sources, including lower power consumption, longer lifetime, increased physical resilience, smaller size, and faster switching [87]. The recombination of electrons and electron holes in a semiconductor creates light (infrared, visible, or UV) in an LED, a process known as "electroluminescence." The energy band gap of the semiconductors used determines the wavelength of light [88]. LEDs come in a variety of packages to suit a variety of uses. For usage as an indication or pilot lamp, a single or a few LED junctions can be packed into a single tiny device [89]. Control circuits for LED arrays can range from a simple resistor to blinking or color changing control, as well as an addressable controller for RGB devices, all of which can be found in the same device [90].

GQDs have piqued interest as prospective luminophores due to a desire to produce new forms of nanomaterials that are free of health and environmental risks but have significant opto-electrical capabilities similar to QDs [91]. Dong et al. [91] by means of DFT used QDs to make a white-light-emitting diode with a brightness of 798 cd m^{-2}. The strain introduced by curvature opens an electronic bandgap of 250 meV in the graphene, and two additional blue emission peaks are observed in the luminescent spectrum of the quantum dot [92]. These additional peaks are the result of a splitting of the graphene's lowest unoccupied orbitals into three separate energy levels, according to density functional theory calculations.

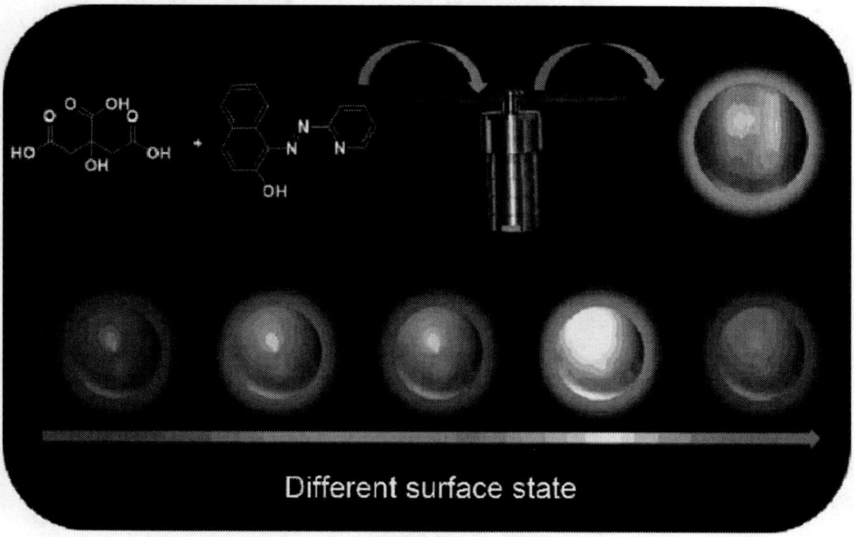

Figure 6. Multicolor carbon dots with concentration-tunable fluorescence and solvent-affected aggregation states for white light-emitting diodes. Reproduced from [93]. Copyright 2022. Springer Nature, Switzerland AG. Part of Springer Nature. https://www.springernature.com/.

QDs are combined with other emissive materials in a multilayer light-emitting diode to produce white emission [87]. Chen et al. [94] reported that GQDs show excellent flexibility and emission property meeting the requirements for the production of white LEDs. Yin et al. [51] fabricated yellowish-green emitting GQDs using a method that is eco-friendly and cost effective (Figure 6). GQDs were cast upon UV-curable adhesive, this formed polymer film of the GQD, having a superior optical transparency and good monodispersing properties. With simple adjustments of the thickness of the GQD phosphors, GQD based white LEDs of different kinds were achieved.

The electroluminescence of GQD-LEDs has been reported to approach 1000 cd m^{-2}. These devices are made possible by a unique synthesis technique that uses solvothermal production of graphite intercalation compounds between graphite powder and sodium potassium tartrate to manufacture GQDs with little oxidation and high quantum yields according to previous research by [20]. Kim et al. [95] confirmed the potential of GQD for the applications of GQD-based LEDs as next-generation real optical sources.

2.6. Sensors

Module equipment with distinctive properties for detection in its environments is classified as sensors [96]. Sensor materials are multidisciplinary such as biosensors, chemical sensors, biochemical sensors, image sensors, monitoring sensors and others.

GQDs have been considered as good sensor materials with regards to their valence and conduction bands which makes them a good semiconductor with zero-bandgap [28]. The fine structure constant determines the optical transmittance of a single monolayer of graphene, and its electronic properties are likely what drew researchers to this material [72]. The spectra of absorption for GQDs show a prominent peak at about 230 nm which has been assigned to the $\pi \rightarrow \pi$ * excitation of the π–bonds [97]. Theoretically, graphene is a perfect atomic monolayer with infinite 2D expansion [98]. GQD is a form of graphene with less than 100 nm with discrete properties due to the confinement which is obvious when size is smaller than 100 nm.

Calculations using density functional theory (DFT) suggest that the bandgap of GQDs increases to approximately 2 eV in a GQD consisting of 20 aromatic rings and 7 eV for benzene [78]. Previous reports on the potential application of GQDs show that these materials are less toxic and more hydrophobic. Also, the intrinsic photoluminescence of GQDs can be used for efficient tracking of human cells in vitro without further modification biomarker dyes [99, 100]. To improve the qualities of GQDs, many approaches have been examined. For instance, Khodadadei et al. [101] synthesized nitrogen-doped GQDs using a hydrothermal method as an efficient anticancer drug delivery system. However, GQDs and carbon-based nano-compounds have been introduced as effective sensors for different gases. With its outstanding properties which include ultrahigh surface to volume ratio and high carrier mobility, GQD have been efficiently used to detect dissolved gases [102].

Transition metal-doped graphene offers the best gas-sensing performance of all the surface modification techniques. Due to its two-dimensional structure, graphene as an array of sp^2 linked carbons holds extraordinary features like a high surface-to-volume fraction and excellent transport capabilities [103]. Reports shows that graphene-like boron carbonitrides (B_x, C_y, & N_z) with huge surface areas ranging from 1500 to 2000m^2/g demonstrate strong CO_2 absorption [104]. However, GQDs modified with certain dopants have great gas sensing capabilities, the dopants improve the selectivity and sensitivity to certain gases in particular [105]. Gui et al. [106] studied the adsorption of C_2H_2 and CH_4 on Mn-doped graphene and found that, with the exception of CH_4, the adsorption energy of gases on Mn-graphene has significantly risen due to electronic hybridization. Amaya-Roncancio et al. [107] studied the adsorption of H_2 on Ni-doped graphene surface. The study revealed that Ni-doped graphene surface had better stability attribute over pristine graphene thus, accounting for its better adsorption properties as observed. Among all surface modifications, transition metal doped graphene has expressly outstanding gas-sensing performance. The investigation of the adsorption behaviour of selected gas molecules on Ag, Au, Pd, and Ga-doped graphene sheet shows a remarkable increase in electrical conductivity pattern of the resultant surfaces [108].

Conclusions, Knowledge Gaps, and Future Prospects

GQDs have made significant progress as arguably the newest ‘luminaries’ of the family of carbon. With their significant development in the last decade, they have drawn attention to themselves as a result of their unique optical and electrical properties. With their unique ability of accommodating dopants and functional groups, their excellence in dispersibility, and non-zero bandgap among others, GQDs have shown remarkable momentum in its application across various areas. We have reviewed the recent progress for their applications in energy and storage devices. It appears that GQDs have improved the specific capacity of Li-ion batteries by a factor of about 1.4, and also resulted in an approximate percentage increase of 15% of their cycling stability.

Their advantage has been outlined in this chapter. One of the major gaps and challenges lies in the area of GQD synthesis. One drawback is the production of GQD from the bottom-up approach, such as low yields as a result on purification challenges in the removal of small organic materials that are unreacted. Various problems include the fact that it has a non-zero bandgap, this also kills the possibility of GQD applications in some optoelectrical devices.

Climate change and economic sustainability is an area that makes energy applications very important, and it is very clear from the reviewed studies that GQDs should be highly considered for wider range application, because despite their drawbacks, they have great future prospects for application in energy and storage devices, because it can be clearly seen from various studies that they increase device performance noticeably.

References

[1] Ren W., Cheng H.-M. The global growth of graphene. *Nature Nanotechnology 2014;9:726–30.*

[2] Tang Q., Zhou Z. Graphene-analogous low-dimensional materials. *Progress in Materials Science* 2013;58:1244–315. https://doi.org/10.1016/j.pmatsci.2013.04.003.

[3] Wu X., Zheng Y., Luo Y., Zhang J., Yi Z., Wu X., et al. A four-band and polarization-independent BDS-based tunable absorber with high refractive index sensitivity. *Physical Chemistry Chemical Physics* 2021;23:26864–73.

[4] King P. J., Khan U., Lotya M., De S., Coleman J. N. Improvement of transparent conducting nanotube films by addition of small quantities of graphene. *Acs Nano* 2010;4:4238–46.

[5] Zurutuza A., Marinelli C. Challenges and opportunities in graphene commercialization. *Nature Nanotechnology* 2014;9:730–4.

[6] Zahir N., Magri P., Luo W., Gaumet J. J., Pierrat P. Recent advances on graphene quantum dots for electrochemical energy storage devices. *Energy & Environmental Materials* 2022;5:201–14.

[7] Wang X., Kalali E. N., Wan J.-T., Wang D.-Y. Carbon-family materials for flame retardant polymeric materials. *Progress in Polymer Science* 2017;69:22–46.

[8] Zdrazil L., Zahradnicek R., Mohan R., Sedlacek P., Nejdl L., Schmiedova V., et al. Preparation of graphene quantum dots through liquid phase exfoliation method. *Journal of Luminescence* 2018;204:203–8.

[9] Zhang Y., Zhang K., Jia K., Liu G., Ren S., Li K., et al. Preparation of coal-based graphene quantum dots/α-Fe2O3 nanocomposites and their lithium-ion storage properties. *Fuel* 2019;241:646–52.

[10] Callendar G. S. On the amount of carbon dioxide in the atmosphere. *Tellus* 1958;10:243–8.

[11] Langenheim J. H. Amber: A Botanical Inquiry: Amber provides an evolutionary framework for interdisciplinary studies of resin-secreting plants. *Science* 1969;163:1157–69.

[12] Padwa A. Photochemistry of the carbon-nitrogen double bond. *ACS Publications* 2002. https://doi.org/10.1021/cr60305a004.

[13] Bruno F., Sciortino A., Buscarino G., Soriano M. L., Ríos Á., Cannas M., et al. A comparative study of top-down and bottom-up carbon nanodots and their interaction with mercury ions. *Nanomaterials* 2021;11:1265.

[14] Kumar N., Salehiyan R., Chauke V., Botlhoko O. J., Setshedi K., Scriba M., et al. Top-down synthesis of graphene: A comprehensive review. *FlatChem* 2021;27:100224.

[15] Peng J., Gao W., Gupta B. K., Liu Z., Romero-Aburto R., Ge L., et al. Graphene quantum dots derived from carbon fibers. *Nano Letters* 2012;12:844–9.

[16] Zhao M. Direct Synthesis of Graphene Quantum Dots with Different Fluorescence Properties by Oxidation of Graphene Oxide Using Nitric Acid. *Applied Sciences* 2018;8:1303. https://doi.org/10.3390/app8081303.

[17] Hassan M., Gomes V. G. Coal derived carbon nanomaterials–Recent advances in synthesis and applications. *Applied Materials Today* 2018;12:342–58.

[18] Ye R., Xiang C., Lin J., Peng Z., Huang K., Yan Z., et al. Coal as an abundant source of graphene quantum dots. *Nature Communications* 2013;4:1–7.

[19] Kaciulis S., Mezzi A., Soltani P., Pizzoferrato R., Ciotta E., Prosposito P. Graphene quantum dots obtained by unfolding fullerene. *Thin Solid Films* 2019;673:19–25.

[20] Song S. H., Jang M.-H., Chung J., Jin S. H., Kim B. H., Hur S.-H., et al. Highly efficient light-emitting diode of graphene quantum dots fabricated from graphite intercalation compounds. *Advanced Optical Materials* 2014;2:1016–23.

[21] Lu L., Zhu Y., Shi C., Pei Y. T. Large-scale synthesis of defect-selective graphene quantum dots by ultrasonic-assisted liquid-phase exfoliation. *Carbon* 2016;109:373–83.

[22] He M., Guo X., Huang J., Shen H., Zeng Q., Wang L. Mass production of tunable multicolor graphene quantum dots from an energy resource of coke by a one-step electrochemical exfoliation. *Carbon* 2018;140:508–20.

[23] Calabro R. L., Yang D.-S., Kim D. Y. Liquid-phase laser ablation synthesis of graphene quantum dots from carbon nano-onions: Comparison with chemical oxidation. *Journal of Colloid and Interface Science* 2018;527:132–40.

[24] Choi Y., Choi Y., Kwon O.-H., Kim B.-S. Carbon Dots: Bottom-Up Syntheses, Properties, and Light-Harvesting Applications. *Chemistry – An Asian Journal* 2018;13:586–98. https://doi.org/10.1002/asia.201701736.

[25] Chen W., Lv G., Hu W., Li D., Chen S., Dai Z. Synthesis and applications of graphene quantum dots: a review. *Nanotechnology Reviews* 2018;7:157–85.

[26] Danial W. H., Norhisham N. A., Ahmad Noorden A. F., Abdul Majid Z., Matsumura K., Iqbal A. A short review on electrochemical exfoliation of graphene and graphene quantum dots. *Carbon Letters* 2021;31:371–88.

[27] Das P., Ganguly S., Banerjee S., Das N. C. Graphene based emergent nanolights: a short review on the synthesis, properties and application. *Res Chem Intermed* 2019;45:3823–53. https://doi.org/10.1007/s11164-019-03823-2.

[28] Ghaffarkhah A., Hosseini E., Kamkar M., Sehat A. A., Dordanihaghighi S., Allahbakhsh A., et al. Synthesis, Applications, and Prospects of Graphene Quantum Dots: A Comprehensive Review. *Small* 2022;18:2102683. https://doi.org/10.1002/smll.202102683.

[29] Sweetman M. J., Hickey S. M., Brooks D. A., Hayball J. D., Plush S. E. A practical guide to prepare and synthetically modify graphene quantum dots. *Advanced Functional Materials* 2019;29:1808740.

[30] Agbo E. P., Edet C. O., Magu T. O., Njok A. O., Ekpo C. M., Louis H. Solar energy: A panacea for the electricity generation crisis in Nigeria. *Heliyon* 2021;7:e07016.

[31] Agbo E. P., Ekpo C. M., Edet C. O. Analysis of the effects of meteorological parameters on radio refractivity, equivalent potential temperature and field strength via Mann-Kendall test. *Theoretical and Applied Climatology* 2021;143:1437–56.

[32] Agbo E. P., Ettah E. B., Eno E. E. The impacts of meteorological parameters on the seasonal, monthly, and annual variation of radio refractivity. *Indian J Phys* 2021;95:195–207. https://doi.org/10.1007/s12648-020-01711-9.

[33] Agbo E. P. The role of statistical methods and tools for weather forecasting and modeling. Weather Forecasting, *IntechOpen;* 2021, p. 3–22.

[34] Agbo E. P., Ekpo C. M. Trend analysis of the variations of ambient temperature using Mann-Kendall test and Sen's estimate in Calabar, southern Nigeria. *Journal of Physics: Conference Series,* vol. 1734, IOP Publishing; 2021, p. 012016.

[35] Alhaji U. U., Yusuf A. S., Edet C. O., Oche C. O., Agbo E. P. *Trend Analysis of Temperature in Gombe State Using Mann Kendall Trend Test. JSRR* 2018;20:1–9. https://doi.org/10.9734/JSRR/2018/42029.

[36] Dincer I. Renewable energy and sustainable development: a crucial review. *Renewable and Sustainable Energy Reviews* 2000;4:157–75.

[37] Gill A. B. Offshore renewable energy: ecological implications of generating electricity in the coastal zone. *Journal of Applied Ecology* 2005:605–15.

[38] Huang S.-C., Lo S.-L., Lin Y.-C. To re-explore the causality between barriers to renewable energy development: A case study of wind energy. *Energies* 2013;6:4465–88.

[39] Ponomarenko L. A., Schedin F., Katsnelson M. I., Yang R., Hill E. W., Novoselov K. S., et al. Chaotic Dirac billiard in graphene quantum dots. *Science* 2008.

[40] Tian P., Tang L., Teng K. S., Lau S. P. Graphene quantum dots from chemistry to applications. *Materials Today Chemistry* 2018;10:221–58.

[41] Huang M., Feng M., Li H., Huang P., Su Q., Zhang F., et al. Rapid microwave-assisted synthesis of SnO2 quantum dots/reduced graphene oxide composite with its application in lithium-ion battery. *Materials Letters* 2017; 209:260–3.

[42] Jansen A. N., Kahaian A. J., Kepler K. D., Nelson P. A., Amine K., Dees D. W., et al. Development of a high-power lithium-ion battery. *Journal of Power Sources* 1999;81:902–5.

[43] Megahed S., Ebner W. Lithium-ion battery for electronic applications. *Journal of Power Sources* 1995;54:155–62.

[44] Pattarapongdilok N., Parasuk V. Adsorptions of lithium ion/atom and packing of Li ions on graphene quantum dots: Application for Li-ion battery. *Computational and Theoretical Chemistry* 2020;1177:112779.

[45] Scrosati B. Recent advances in lithium ion battery materials. *Electrochimica Acta* 2000;45:2461–6.

[46] Scrosati B. History of lithium batteries. *Journal of Solid State Electrochemistry* 2011;15:1623–30.

[47] Tarascon J.-M., Armand M. Issues and challenges facing rechargeable lithium batteries. Materials for sustainable energy: a collection of peer-reviewed research and review articles from Nature Publishing Group, World Scientific; 2011, p. 171–9.

[48] Gao L., Wu G., Ma J., Jiang T., Chang B., Huang Y., et al. SnO2 quantum dots@ graphene framework as a high-performance flexible anode electrode for lithium-ion batteries. *ACS Applied Materials & Interfaces* 2020;12:12982–9.

[49] Lijuan K., Yongqiang Y., Ruiyi L., Zaijun L. Phenylalanine-functionalized graphene quantum dot-silicon nanoparticle composite as an anode material for lithium ion batteries with largely enhanced electrochemical performance. *Electrochimica Acta* 2016;198:144–55.

[50] Ruiyi L., Yuanyuan J., Xiaoyan Z., Zaijun L., Zhiguo G., Guangli W., et al. Significantly enhanced electrochemical performance of lithium titanate anode for lithium ion battery by the hybrid of nitrogen and sulfur co-doped graphene quantum dots. *Electrochimica Acta* 2015;178:303–11.

[51] Yin L., Zhou J., Li W., Zhang J., Wang L. Yellow fluorescent graphene quantum dots as a phosphor for white tunable light-emitting diodes. *RSC Adv* 2019;9:9301–7. https://doi.org/10.1039/C8RA10353D.

[52] Harismah K., Mirzaei M., Moradi R. DFT studies of single lithium adsorption on coronene. *Zeitschrift Für Naturforschung A* 2018;73:685–91.

[53] Papanek P., Radosavljević M., Fischer J. E. Lithium insertion in disordered carbon-hydrogen alloys: intercalation vs covalent binding. *Chemistry of Materials* 1996;8:1519–26.

[54] Zhao X., Wu Y., Wang Y., Wu H., Yang Y., Wang Z., et al. High-performance Li-ion batteries based on graphene quantum dot wrapped carbon nanotube hybrid anodes. *Nano Res* 2020;13:1044–52. https://doi.org/10.1007/s12274-020-2741-9.

[55] Fan X., Zheng W. T., Kuo J.-L., Singh D. J. Adsorption of single Li and the formation of small Li clusters on graphene for the anode of lithium-ion batteries. *ACS Applied Materials & Interfaces* 2013;5:7793–7.

[56] He J., Chen Y., Li P., Fu F., Wang Z., Zhang W. Self-assembled CoS2 nanoparticles wrapped by CoS2-quantum-dots-anchored graphene nanosheets as superior-capability anode for lithium-ion batteries. *Electrochimica Acta* 2015;182:424–9.

[57] Chen L., Chen W.-L., Wang X.-L., Li Y.-G., Su Z.-M., Wang E.-B. Polyoxometalates in dye-sensitized solar cells. *Chemical Society Reviews* 2019;48:260–84.

[58] Chen Z., Hou J., Liu Q., Zhou Q., Liu H., Xu C. Graphene quantum dots modified nanoporous SiAl composite as an advanced anode for lithium storage. *Electrochimica Acta* 2019;318:228–35.

[59] Chao D., Zhu C., Xia X., Liu J., Zhang X., Wang J., et al. Graphene quantum dots coated VO2 arrays for highly durable electrodes for Li and Na ion batteries. *Nano Letters* 2015;15:565–73.

[60] Kumar D. K., Suazo-Davila D., García-Torres D., Cook N. P., Ivaturi A., Hsu M.-H., et al. Low-temperature titania-graphene quantum dots paste for flexible dye-sensitised solar cell applications. *Electrochimica Acta* 2019;305:278–84.

[61] Teymourinia H., Salavati-Niasari M., Amiri O., Farangi M. Facile synthesis of graphene quantum dots from corn powder and their application as down conversion effect in quantum dot-dye-sensitized solar cell. *Journal of Molecular Liquids* 2018;251:267–72.

[62] Yang H. B., Dong Y. Q., Wang X., Khoo S. Y., Liu B., Li C. M. Graphene quantum dots-incorporated cathode buffer for improvement of inverted polymer solar cells. *Solar Energy Materials and Solar Cells* 2013;117:214–8.

[63] Jahantigh F., Ghorashi S. B., Bayat A. Hybrid dye sensitized solar cell based on single layer graphene quantum dots. *Dyes and Pigments* 2020;175:108118.

[64] Majumder T., Dhar S., Chakraborty P., Debnath K., Mondal S. P. S., N Co-Doped Graphene Quantum Dots Decorated C-Doped ZnO Nanotaper Photoanodes for Solar Cells Applications. *NANO* 2019;14:1950012. https://doi.org/10.1142/S1793292019500127.

[65] Sharma V., Jha P. K. Enhancement in power conversion efficiency of edge-functionalized graphene quantum dot through adatoms for solar cell applications. *Solar Energy Materials and Solar Cells* 2019;200:109908.

[66] Majumder T., Dhar S., Chakraborty P., Debnath K., Mondal S. P. Advantages of ZnO nanotaper photoanodes in photoelectrochemical cells and graphene quantum dot sensitized solar cell applications. *Journal of Electroanalytical Chemistry* 2018;813:92–101.

[67] Liang J., Liu Y., Si Z., Wei G., Weng D., Kang F. Graphene quantum dots piecing together into graphene on nano Au for overall water splitting. *Carbon* 2021;178:265–72.

[68] Xue F., Si Y., Wang M., Liu M., Guo L. Toward efficient photocatalytic pure water splitting for simultaneous H2 and H2O2 production. *Nano Energy* 2019;62:823–31.

[69] Calabro R. L., Yang D.-S., Kim D. Y. Controlled Nitrogen Doping of Graphene Quantum Dots through Laser Ablation in Aqueous Solutions for Photoluminescence and Electrocatalytic Applications. *ACS Appl Nano Mater* 2019;2:6948–59. https://doi.org/10.1021/acsanm.9b01433.

[70] Yuan A., Lei H., Xi F., Liu J., Qin L., Chen Z, et al. Graphene quantum dots decorated graphitic carbon nitride nanorods for photocatalytic removal of antibiotics. *Journal of Colloid and Interface Science* 2019;548:56–65.

[71] Rahimi K., Yazdani A., Ahmadirad M. Facile preparation of zinc oxide nanorods surrounded by graphene quantum dots both synthesized via separate pyrolysis procedures for photocatalyst application. *Materials Research Bulletin* 2018;98:148–54.

[72] Rahimi K., Yazdani A., Ahmadirad M. Graphene quantum dots enhance UV photoresponsivity and surface-related sensing speed of zinc oxide nanorod thin films. *Materials & Design* 2018;140:222–30.

[73] Fan J., Li D., Wang X. Effect of modified graphene quantum dots on photocatalytic degradation property. *Diamond and Related Materials* 2016;69:81–5.

[74] Khan M. S., Riaz N., Shaikh A. J., Shah J. A., Hussain J., Irshad M., et al. Graphene quantum dot and iron co-doped TiO2 photocatalysts: Synthesis, performance evaluation and phytotoxicity studies. *Ecotoxicology and Environmental Safety* 2021;226:112855.

[75] Kumar S., Dhiman A., Sudhagar P., Krishnan V. ZnO-graphene quantum dots heterojunctions for natural sunlight-driven photocatalytic environmental remediation. *Applied Surface Science* 2018;447:802–15.

[76] Xie H., Hou C., Wang H., Zhang Q., Li Y. S., N co-doped graphene quantum dot/TiO2 composites for efficient photocatalytic hydrogen generation. *Nanoscale Research Letters* 2017;12:1–8.

[77] Hadjipaschalis I., Poullikkas A., Efthimiou V. Overview of current and future energy storage technologies for electric power applications. *Renewable and Sustainable Energy Reviews* 2009;13:1513–22.

[78] Bacon M., Bradley S. J., Nann T. Graphene quantum dots. *Particle & Particle Systems Characterization* 2014;31:415–28.

[79] Sun H., Wu L., Wei W., Qu X. Recent advances in graphene quantum dots for sensing. *Materials Today* 2013;16:433–42.

[80] Jayalakshmi M., Balasubramanian K. Simple capacitors to supercapacitors-an overview. *Int J Electrochem Sci* 2008;3:1196–217.

[81] Zhang R., Zhou X. Y., Zhang D., Lou W. K., Zhai F., Chang K. Electronic and magneto-optical properties of monolayer phosphorene quantum dots. *2D Materials* 2015;2:045012.

[82] Shaker M., Riahifar R., Li Y. A review on the superb contribution of carbon and graphene quantum dots to electrochemical capacitors' performance: Synthesis and application. *FlatChem* 2020;22:100171. https://doi.org/10.1016/j.flatc.2020.100171.

[83] Aboutalebi S. H., Jalili R., Esrafilzadeh D., Salari M., Gholamvand Z., Aminorroaya Yamini S, et al. High-performance multifunctional graphene yarns: toward wearable all-carbon energy storage textiles. *ACS Nano* 2014;8:2456–66.

[84] Zhang S., Sui L., Dong H., He W., Dong L., Yu L. High-performance supercapacitor of graphene quantum dots with uniform sizes. *ACS Applied Materials & Interfaces* 2018;10:12983–91.

[85] Liu W.-W., Feng Y.-Q., Yan X.-B., Chen J.-T., Xue Q.-J. Superior micro-supercapacitors based on graphene quantum dots. *Advanced Functional Materials* 2013;23:4111–22.

[86] Li L.-L., Ji J., Fei R., Wang C.-Z., Lu Q, Zhang J.-R., et al. A facile microwave avenue to electrochemiluminescent two-color graphene quantum dots. *Advanced Functional Materials* 2012;22:2971–9.

[87] Nair G. B., Swart H. C., Dhoble S. J. A review on the advancements in phosphor-converted light emitting diodes (pc-LEDs): Phosphor synthesis, device fabrication and characterization. *Progress in Materials Science* 2020;109:100622.

[88] Muthu S., Schuurmans F. J., Pashley M. D. Red, green, and blue LEDs for white light illumination. *IEEE Journal of Selected Topics in Quantum Electronics* 2002;8:333–8.

[89] Held G. *Introduction to light emitting diode technology and applications*. Auerbach publications; 2016.

[90] Lai C.-C., Huang M.-J., Chou H.-H., Liao C.-Y., Rajamalli P., Cheng C.-H. m-Indolocarbazole Derivative as a Universal Host Material for RGB and White Phosphorescent OLEDs. *Advanced Functional Materials* 2015;25:5548–56.

[91] Dong Y., Wang Y.-K., Yuan F., Johnston A., Liu Y., Ma D, et al. Bipolar-shell resurfacing for blue LEDs based on strongly confined perovskite quantum dots. *Nature Nanotechnology* 2020;15:668–74.

[92] Son D. I., Kwon B. W., Park D. H., Seo W.-S., Yi Y., Angadi B, et al. Emissive ZnO–graphene quantum dots for white-light-emitting diodes. *Nature Nanotechnology* 2012;7:465–71.

[93] Yan F., Jiang Y., Sun X., Wei J., Chen L., Zhang Y. Multicolor carbon dots with concentration-tunable fluorescence and solvent-affected aggregation states for white light-emitting diodes. *Nano Res* 2020;13:52–60. https://doi.org/10.1007/s12274-019-2569-3.

[94] Chen Y.-X., Lu D., Wang G.-G., Huangfu J., Wu Q.-B., Wang X.-F., et al. Highly efficient orange emissive graphene quantum dots prepared by acid-free method for white LEDs. *ACS Sustainable Chemistry & Engineering* 2020;8:6657–66.

[95] Kim D. H., Kim T. W. Ultrahigh current efficiency of light-emitting devices based on octadecylamine-graphene quantum dots. *Nano Energy* 2017;32:441–7.

[96] Vetelino J., Reghu A. *Introduction to sensors*. CRC press; 2017.

[97] Fan Z., Li Y., Li X., Fan L., Zhou S., Fang D., et al. Surrounding media sensitive photoluminescence of boron-doped graphene quantum dots for highly fluorescent dyed crystals, chemical sensing and bioimaging. *Carbon* 2014;70:149–56.

[98] Chung S., Revia R. A., Zhang M. Graphene quantum dots and their applications in bioimaging, biosensing, and therapy. *Advanced Materials* 2021;33:1904362.

[99] Kumar P., Dhand C., Dwivedi N., Singh S., Khan R., Verma S., et al. Graphene quantum dots: A contemporary perspective on scope, opportunities, and sustainability. *Renewable and Sustainable Energy Reviews* 2022;157:111993.

[100] Schroeder K. L., Goreham R. V., Nann T. Graphene quantum dots for theranostics and bioimaging. *Pharmaceutical Research* 2016;33:2337–57.

[101] Khodadadei F., Safarian S., Ghanbari N. Methotrexate-loaded nitrogen-doped graphene quantum dots nanocarriers as an efficient anticancer drug delivery system. *Materials Science and Engineering: C* 2017;79:280–5.

[102] Chen F., Gao W., Qiu X., Zhang H., Liu L., Liao P., et al. Graphene quantum dots in biomedical applications: recent advances and future challenges. *Frontiers in Laboratory Medicine* 2017;1:192–9.

[103] Toh R. J., Poh H. L., Sofer Z., Pumera M. Transition Metal (Mn, Fe, Co, Ni)-Doped Graphene Hybrids for Electrocatalysis. *Chemistry–An Asian Journal* 2013;8:1295–300.

[104] Beniwal S., Hooper J., Miller D. P., Costa P. S., Chen G., Liu S.-Y., et al. Graphene-like boron–carbon–nitrogen monolayers. *ACS Nano* 2017;11:2486–93.

[105] Kaur M., Kaur M., Sharma V. K. Nitrogen-doped graphene and graphene quantum dots: A review onsynthesis and applications in energy, sensors and environment. *Advances in Colloid and Interface Science* 2018;259:44–64.

[106] Gui Y., Peng X., Liu K., Ding Z. Adsorption of C2H2, CH4 and CO on Mn-doped graphene: Atomic, electronic, and gas-sensing properties. *Physica E: Low-Dimensional Systems and Nanostructures* 2020;119:113959.

[107] Amaya-Roncancio S., Blanco A. G., Linares D. H., Sapag K. DFT study of hydrogen adsorption on Ni/graphene. *Applied Surface Science* 2018;447:254–60.

[108] Tang X., Du A., Kou L. Gas sensing and capturing based on two-dimensional layered materials: Overview from theoretical perspective. *Wiley Interdisciplinary Reviews: Computational Molecular Science* 2018;8:e1361.

Chapter 5

Graphene Quantum Dots: Promising Catalysts for Electrocatalytic Water Splitting

Samika Anand
R. Madhushree
U. C. Jadan Resnik Jaleel
Dephan Pinheiro
and K. R. Sunaja Devi*
Department of Chemistry, CHRIST (Deemed to be University), Bengaluru, India

Abstract

In recent times, graphene quantum dots have attracted attention from both academia and industry due to their low cost, non-toxicity, abundance, inertness, stability, photoluminescence, and ease of functionalization. Graphene quantum dots are doped with heteroatoms and modified structurally with metals/metal oxides to form composites that find immense application in catalysis. Electrocatalytic water splitting is an energy efficient route for oxygen evolution, hydrogen evolution, and oxygen reduction reactions. Graphene quantum dots based composites have been extensively employed for electrocatalytic water splitting applications, with remarkable results. This chapter focuses on electrocatalytic applications of modified graphene quantum dots in water splitting.

Keywords: graphene quantum dots, water splitting, hydrogen evolution reaction, oxygen evolution reaction, oxygen reduction reaction

* Corresponding Author's Email: sunajadevi.kr@christuniversity.in.

In: Graphene Quantum Dots (GQDs)
Editor: Aleksey Kuznetsov
ISBN: 979-8-89113-638-0

1. Introduction

Sustainable use of energy is one of the biggest challenges of the 21st century. The recent technologies involve the development of hydrogen energy from electrocatalytic as well as photocatalytic water splitting and metal air batteries [1]. Overall water splitting (OWS) produces hydrogen and oxygen via Hydrogen evolution reaction (HER) and Oxygen evolution reaction (OER) at the cathode and anode, respectively [2]. Oxygen reduction reaction (ORR) is typically the reverse of OER, and results in the reduction of oxygen to give water. These reactions, however, suffer from sluggish kinetics and hence require the use of electro-active catalysts to improve the electron transfer process[3]. An efficient catalyst will not only reduce the activation energy barrier but also remarkably quicken the process, in turn simplifying the mechanism [3].

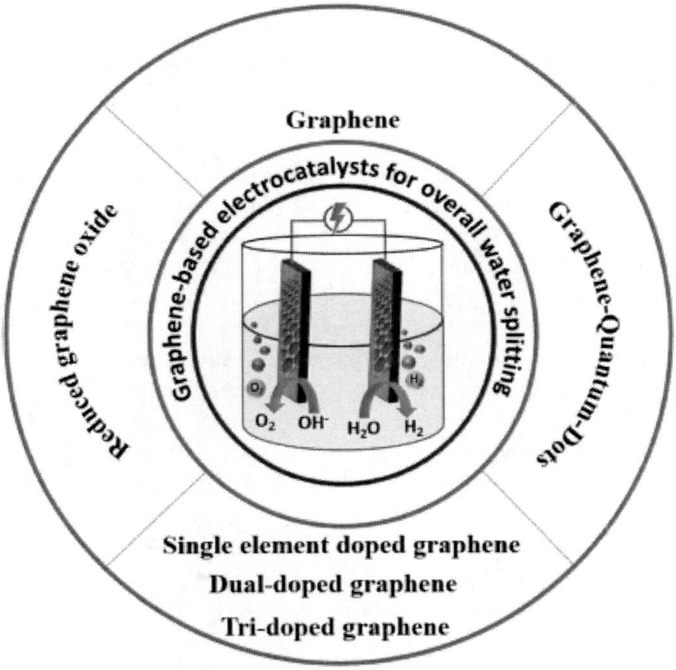

Figure 1. Schematic representation of GQDs based electrocatalysts for OWS[2]. Reprinted with permission from ref[2]. Copyright 2020 Springer.

Conventional electrocatalysts include noble-metal catalysts like Pt, Ru, Ir, etc., which are expensive and rare [2]. However, a quest for abundant and inexpensive catalysts has been a significant concern for a few decades now. Alternatives like light-induced OWS and the use of carbon-based composites as electrocatalysts have been explored in recent times [4]. Graphene-based materials, with their fascinating properties like electrical conductivity, mechanical strength, large surface area, and low cost, have attracted vast attention from the scientific community. Several reports from the literature talk about the electroactivity of graphene-based materials (Figure 1) [2]. One among them is graphene quantum dots (GQDs), having a lateral size of less than 10 nm and thickness ranging between 0.4-2.0 nm [5]. The quantum confinement and edge effects bestow GQDs with peculiar properties like high surface area, high electron mobility, non-zero energy gap, and photoluminescence [1]. These physicochemical properties make GQDs based materials highly desirable candidates for energy storage and catalysis [5]. Despite several interesting properties, pristine GQDs generally do not give extraordinary results in electrocatalysis. They are doped with heteroatoms like nitrogen, sulphur, phosphorous, and boron to form doped GQDs [6, 7] or combined with metals to form GQDs based composites [8, 9]. Such materials have been reported to show excellent activity in electrochemistry, especially electrocatalytic OWS. This chapter mainly covers the utility of GQDs based materials in electrocatalytic water splitting.

2. Electrocatalytic Water Splitting

Electrocatalytic OWS comprises mainly two reactions, OER and HER, at the anode and cathode, respectively. However, ORR involves reduction of oxygen to give back water. The involvement of electro active materials can result in remarkable hydrogen or oxygen evolution. Some exceptional materials may give good efficiency for both, HER and OER. The following sections give a detailed account of GQDs based composites in HER, OER, and ORR.

2.1. Hydrogen Evolution Reaction (HER)

Recently, carbon-based materials have been employed as electrocatalysts for reactions such as HER and OWS. Some of these materials include graphene, carbon nanotubes, and carbon quantum dots [10]. In particular, graphene and

GQDs have attracted interest due to their properties like low cost, abundance, stability, excellent charge transfer capabilities, high specific surface areas, high electric conductivities, and environmentally friendly as an electrocatalyst, and its electrocatalytic performances have been enhanced by doping with metals, metal oxides, chalcogenides and non-metals [11]. HER proceeds through a Volmer-Heyrovsky or Volmer-Tafel mechanism and produces H_2 [12]. The HER plays a crucial role in electrochemistry for the generation/production of hydrogen and are pH controlled/dependent. Eq. (1) to (6) represent the H_2 production mechanism. Eq. (1) to Eq. (3) involves the steps under acidic condition, where fast proton transfer and high H_2 production is reported [13]. Whereas Eq. (4) to Eq. (6) are for alkaline conditions, but the HER is very slow as step 4 involves the dissociation of the water molecule.

Under acidic condition

$$H^+ + e^- \rightarrow H_{ads}\ Volmer\ step \tag{1}$$

$$H_{ads} + H_{ads} \rightarrow H_2\ Tafel\ step \tag{2}$$

$$H_{ads} + H^+ + e^- \rightarrow H_2\ Heyrovsky\ step \tag{3}$$

Under alkaline condition

$$H_2O + e^- \rightarrow H_{ads} + OH^-\ Volmer\ step \tag{4}$$

$$H_{ads} + H_{ads} \rightarrow H_2\ Tafel\ step \tag{5}$$

$$H_{ads} + H_2O + e^- \rightarrow H_2 + OH^-\ Heyrovsky\ step \tag{6}$$

Conventional electrocatalysts such as Pt, Ir, or Ru are rare, expensive, and less stable for OWS [14]. GQDs are employed as a substitute for noble metals and metal-free co-catalyst for processes that produce hydrogen [13]. The quantum dots (QDs) can be doped with metals like Ag, Au, Pt, Pd, Cu, and Ni as co-catalyst to effectively increase H_2 generation [15]. The GQDs doped with heteroatoms like nitrogen [16, 17], boron, sulphur, or metal oxides [18, 19] can successfully control the electronic characteristics to enhance the electrocatalytic performance [13].

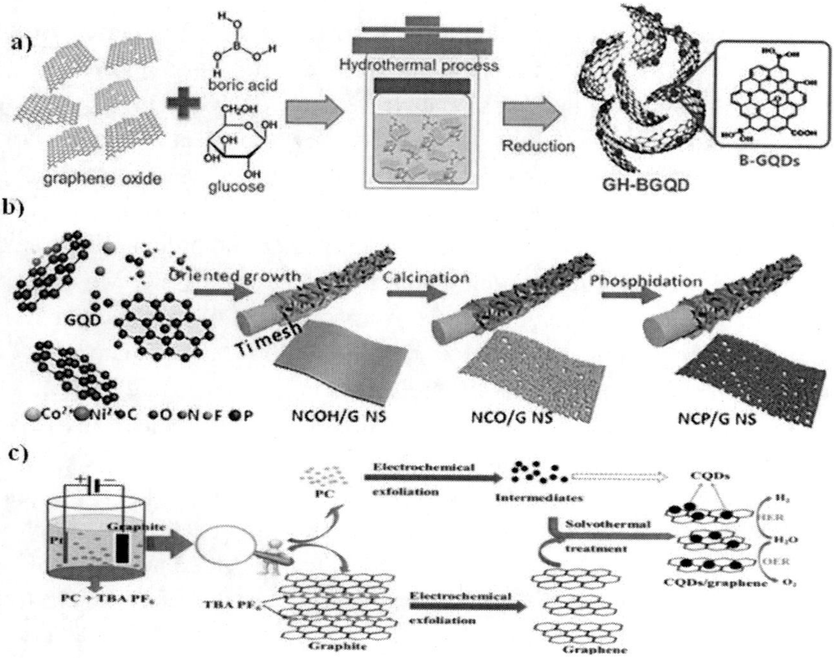

Figure 2. Schematic representation of (a) hydrothermal approach for GH-BGQDs [20]. Reprinted with permission from ref. [20]. Copyright 2019 Willey, (b) fabrication process for NCP/GQD NSs [21]. Reprinted with permission from ref. [21]. Copyright 2018 Elsevier, (c) synthesis route for CQD/GQDs [22]. Reprinted with permission from ref. [22]. Copyright 2019 Royal Society of Chemistry.

Tam et al. prepared a 3D structured graphene hydrogel B-doped GQDs (GH B-GQDs) via hydrothermal approach as depicted in Figure 2a with a Tafel slope of 95 mV dec^{-1} for HER [20]. Tian et al. synthesized nickel-cobalt-phosphide doped GQD nanosheets (NCP/GQD NSs) via a simple fabrication process (Figure 2b). It showed a current density of 100 mA cm^{-2} at an overpotential of 119 mV [21]. Zhao et al. synthesized carbon quantum dots CQDs/GQDs in situ by a two-step procedure, as shown in Figure 2c for the HER that resulted in a Tafel slope value of 80 mV dec^{-1} with an overpotential 194 mV at a current density of 10 mA cm^{-2} [22].

Guo et al. synthesized a 2D - zero gap material electrocatalyst, a coral MoS_2/GQDs via chemical vapour deposition (CVD) method for high hydrogen production with a Tafel slope of 40 mV dec^{-1} and long-term electrocatalytic stability of 2000 cycles. The polarization curves (Figure 3a and 3b), depict that the overpotential of coral MoS_2/GQDs is smaller than the other systems at a current density of 10 mA cm^{-2}, and smaller Tafel slow indicates that hydrogen production is via a combined Volmer-Heyrovsky mechanism [23]. Lv et al. fabricated an electrocatalyst as an electrode by doping N and Ni_3S_2 with GQDs on a Ni foam (Ni_3S_2/NGQDs/NF) for direct water splitting reaction. Figures 3c and 3d show the onset overpotentials of Ni_3S_2/NGQDs/NF and other as-prepared catalysts, with the ternary composite showing an overpotential of 218 mV at a current density of 10 mA cm^{-2}. The lower Tofel slope indicates the fast charge transfer and higher HER kinetics [24]. Li et al. synthesized a noble-metal free electrocatalyst G-Mo-Ni_3S_2 via hydrothermal approach for the lowest overpotential of 68 mV dec^{-1} at 10 mA cm^{-2} with significant stability of more than 50 h. The corresponding polarization curves and Tofel plots are shown in Figures 3e and 3f [25].

Gong et al. reported superior electrocatalytic performance in HER by utilizing GQDs/MoS_2 synthesized via exfoliation, carrying a current density of 10 mA cm^{-2} with a slight Tafel slope of 5.9 mV dec^{-1} [19]. Tsai et al. synthesized CdSe/GQDs via hydrothermal approach for improved photoelectrochemical hydrogen production, using 0.1 M Na_2S/Na_2SO_3 as electrolyte, Pt as a counter electrode, and Ag/AgCl as the reference electrode. Here, the QDs were modified to reduced graphene oxide (RGO) and doped with the CdSe nanoparticles [26].

Sim et al. prepared an attention-grabbing electrocatalyst by co-doping N and F with GQDs that resulted in N, F-GQDs via two-step functionalizing method. Due to the significant electronegativity difference between carbon and fluorine, the catalyst exhibited remarkably high electrochemical reaction properties for HER [27]. Wang et al. established an inexpensive electrocatalyst CoP/G/GQDs, a self-assembled graphene with GQDs doped with CoP, exhibiting a high HER with a Tafel slope of 42.6 mV dec^{-1} and an overpotential of 91.3 mV [28]. Figure 4 depicts the TEM and HRTEM of CoP/G/GQDs. Figures 4a and 4b clearly indicate the uniformly dispersed CoP NPs stacked on graphene/GQDs at a magnification of 500 nm and 100 nm, respectively. Figures 4c and 4d show the lattice fringes with a width of 0.188 nm, which refers to the (211) plane of Co [28].

Riyajuddin et al. synthesized graphene-carbon nanotubes co-doped Ni_2P and CuP_2. The catalyst showed an ultra-low overpotential of 12 mV for HER at current densities 10 mA cm^{-2} and 20 mA cm^{-2} in acidic and alkaline medium [3]. Feng et al. fabricated an allotrope of carbon as QD, fullerene QDs doped with CoNi-LDH/NF showing superior electrolytic activity for HER [29].

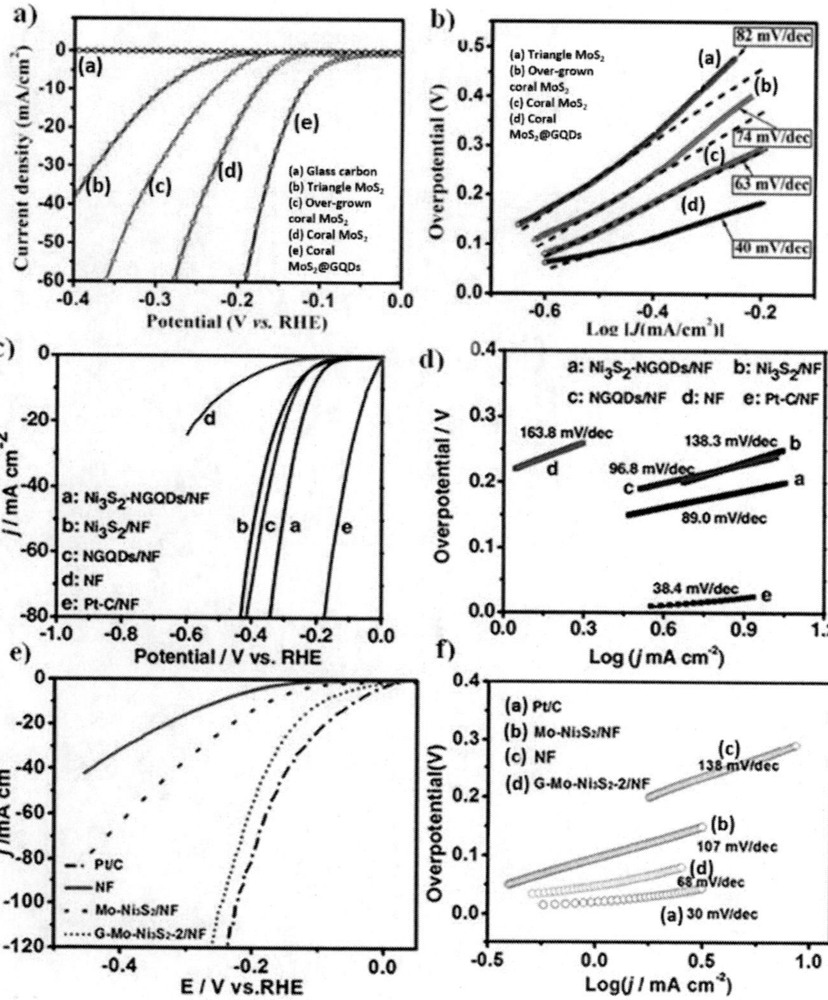

Figure 3. Polarization curves and Tafel plots for HER (a,b) coral MoS_2/GQDs and other as-prepared systems [23]. Reprinted with permission from ref. [23]. Copyright 2017 American Chemical Society. (c,d) Ni_3S_2/NGQDs/NF and other as-prepared systems for HER [24]. Reprinted with permission from ref. [24]. Copyright 2017 Willey. (e,f) G-Mo-Ni_3S_2 polarization curves and Tafel plots with other as-prepared systems for HER [25]. Reprinted with permission from ref. [25]. Copyright 2019 Elsevier.

Xue et al. synthesized a new MXene based GQDs type of electrocatalyst 0D/2D GQDs/$Ti_3C_2T_x$ via etching and solvothermal method, as depicted in Figure 5a. The FE-SEM, TEM, and HR-TEM characterization shows the ideal growth with abundant binding sites for immobilizing tiny GDQs and crystal plane spacing of GQDs as depicted in Figures 5b to 5g. The LSV and Tafel plots are shown in Figures 5h and 5i, respectively, which indicate the lowest onset potential of 130 mV and the lower Tafel slope of 89 mV dec^{-1} for GQDs/$Ti_3C_2T_x$ [30].

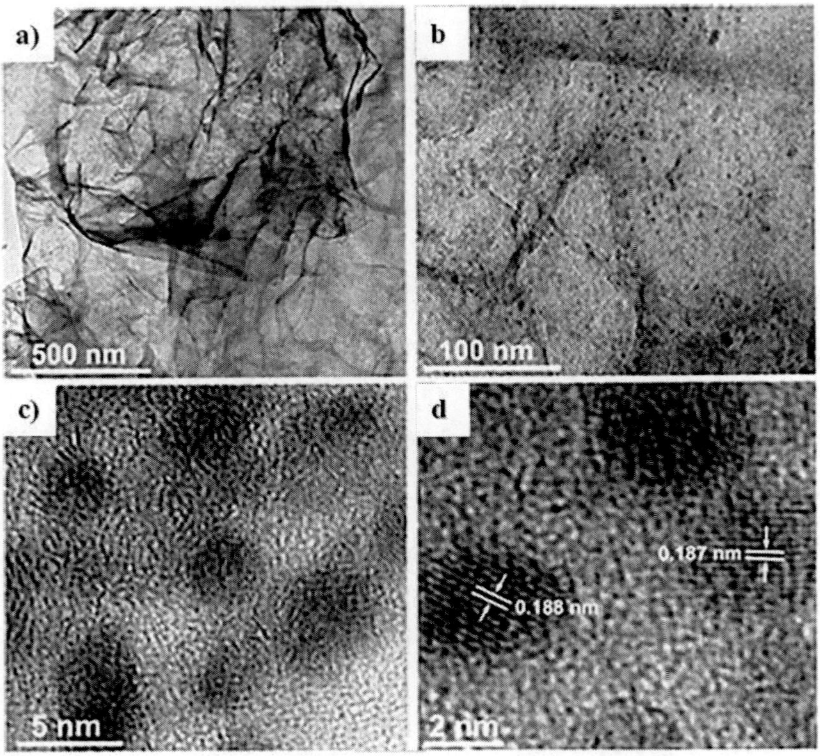

Figure 4. TEM and HRTEM images of CoP/G/GQDs [28]. Reprinted with permission from ref. [28]. Copyright 2017 Willey.

Cao et al. had put forward a bi-functional electrocatalyst, N-doped GQDs sandwiched between LaSrCoO/$MoSe_2$, and the resulting LSC-N-GQDs/ $MoSe_2$ composite had a small overpotential, Tafel slope (64 mV dec^{-1}) and charge transfer resistance in alkaline medium for HER [31]. Gong et al. intercalated MoS_2 with GQDs and doped with Co resulting in GQDs/Co-MoS_2 via hydrothermal approach for enhanced HER in an alkaline medium, with the Tafel slope of 53.8 mV dec^{-1}, which is slightly higher than Pt/C [32]. Cheng et al. have done similar work, where GQDs acted as the carbon support in Pt NPs/C. A remarkable increase in HER was noticed in both acid and alkaline mediums with overpotentials and Tafel slopes of 84 mV, 45 mV dec^{-1} and 300 mV, 152 mV dec^{-1}, respectively [33]. Sharma et al. synthesized various electrocatalysts using oxygen, borinic acid, and boronic acids resulting in H_2-O-GQDs, H_2-BCO_2-GQDs, and H_2-BC_2O-GQDs for HER and studied via density functional theory (DFT) comparing Gibbs energy [34].

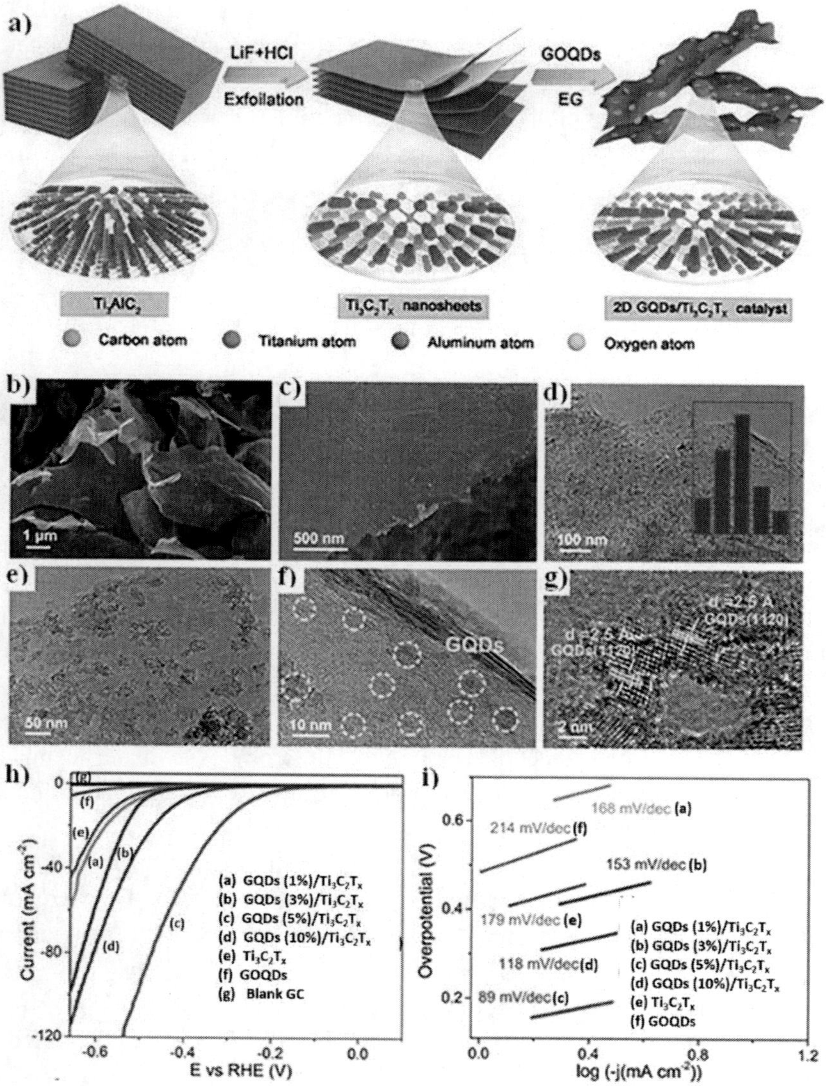

Figure 5. (a) Schematic representation of 0D/2D GQDs/$Ti_3C_2T_x$, (b-c) FE-SEM, (d-e) TEM, (f-g) HR-TEM of 0D/2D GQDs/$Ti_3C_2T_x$, (h-i) LSV curves and Tafel plots of GQDs/$Ti_3C_2T_x$ [30]. Reprinted with permission from ref. [30]. Copyright 2022 Elsevier.

2.2. Oxygen Evolution Reaction (OER)

The production of oxygen by various chemical means, such as oxidation of water or OWS is known as oxygen evolution reaction [35]. Similar to HER, in OER, the measurement of O_2 produced is the basic principle [36]. OER can be measured by using Tafel plots and Tafel slopes. Carbon based materials have shown very good efficacy in OER, HER, and for ORR. Fabricated GQDs, as well as metal doped GQDs, are more efficient in OER.

In OWS, water splits into oxygen (O_2) and hydrogen (H_2). The reaction in which O_2 is produced is shown in Eq. (7) to (10) [37].

$$H_2O(l) + * \rightarrow * OH + H^+ + e^- \tag{7}$$

$$* OH \rightarrow * + H^+ + e^- \tag{8}$$

$$* O + H_2O(l) \rightarrow * OOH + H^+ + e^- \tag{9}$$

$$* OOH \rightarrow O_2(g) + H^+ + e^- \tag{10}$$

Cirone et al. [38] synthesized GQD with Co using a green synthesis method. The Co/GQD nanocomposite showed a very high current density of 43.16 mA cm^{-2}, while Co nanoparticles had a lower current density (21.1 mA cm^{-2}) as compared to Pt/C commercial catalyst (5.99 mA cm^{-2}). Additionally, compared to the other two catalysts under study, the composite material only needed an overpotential of 0.49 V at 10 mA cm^{-2}. According to electrochemical impedance measurements, the composite material had the greatest efficiency for OER in comparison to all the materials examined at various potentials [38].

Tiang and co-workers engineered GQDs with Ni-Co phosphide (NCP/G NS) [21]. The bifunctional bimetallic phosphide based GQDs were more efficient than their oxide equivalents. An electrolyzer with two identical NCP/G NS electrodes showed good performance in OWS (10 mA cm^{-2} at 1.61 V) and higher stability with a catalyst loading of 0.31 mg cm^{-2}. The higher activity of NCP/G NS as compared to $NiCo_2P_2$ nanowires demonstrated the role of GQDs in controlling the morphology thereby enhancing the performance of the catalyst. Bimetallic phosphides have an advantage over their oxide equivalents, according to performance comparisons with $NiCo_2O_4$ and GQD hybrid (NCO/G NSs). The synthesis of GQD, which is doped with Ni and Co phosphides, is shown in Figure 6 [21].

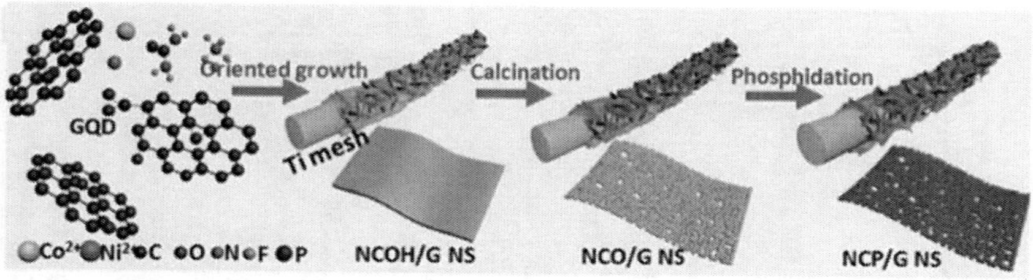

Figure 6. The method of preparation of the NCOH/G NS, NCO/G NS, and NCP/G NSs [21]. Reprinted with permission from ref. [21]. Copyright 2018 Elsevier.

Alves et al. [39] synthesized graphene TiO_2 quantum dots and studied the photoelectrochemical water splitting. In that, they found that the catalysts show more activity in OER. In this study, it was reported that an oleo thermal redox process was used to produce graphene-TiO_2-quantum dots. Nanodroplets of stable oil-water mixture were created using nano-emulsification principles. This emulsion was processed in an oil-based medium at various temperatures. The use of water as a solvent encourages the aggregation of hydrophilic nanoparticles. The cluster of hydrated titanium is generated within the graphene oxide nanodroplet in the oil medium. This feature ensures a tight link between titanium ions and graphene layers, resulting in distributed quantum dots with the smallest possible size. The degree of functionalization of graphene layers and, therefore, the property of photoluminescence is influenced by the temperature and duration of the oleo thermal process. The electrophoresis was used to create a clear thin film for water splitting characterization. It was easy to connect the produced photocurrents and oleo thermal conditions to the level of oxygen functionalization of the graphene layers.

Dong et al. synthesized nitrogen doped Graphene quantum dots (N-GQDs) anchored on NiFe layered double hydroxides sheets (N-GQDs/NiFe-LDH) via a straightforward hydrothermal approach [40]. The catalyst having a hierarchical nanosheet structure (10-180 nm) showed a higher surface area and a range of pore sizes. While examining the OER catalytic activity, a minimal overpotential of 279 and 300 mV at current densities of 20 and 50 mA cm^{-2}, respectively, was seen. The catalyst also demonstrated a low Tafel slope of 47 mV dec^{-1} and high stability (Figures 7a, 7b) [40].

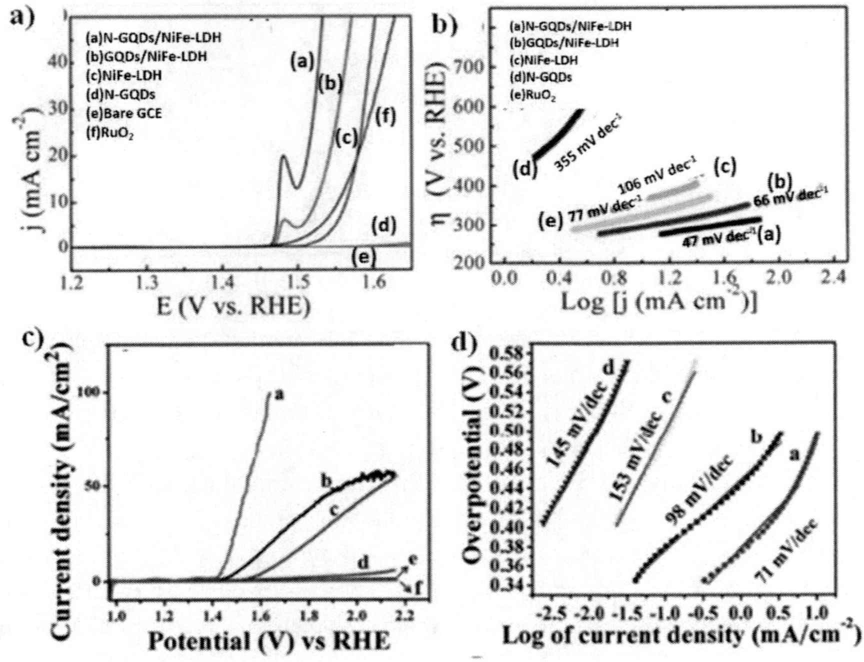

Figure 7. (a) OER Polarization curves and (b) Tafel plot of NGQDs/NiFe-LDH and other as prepared catalysts [40]. Reprinted with permission from ref. [40]. Copyright 2020 Royal Society of Chemistry. (c) LSV curves and (d) Tafel plot of N-GQDs/Co_3O_4 [8]. Reprinted with permission from ref. [8]. Copyright 2017 Elsevier.

Muthurasu and colleagues [8] fabricated bifunctional N-GQDs/Co_3O_4 composite for OER and ORR. The N-GQDs/Co_3O_4 composite showed a greater electrocatalytic activity towards ORR when compared to pure Co_3O_4 and N-GQDs taken as independent entities. However, in OER, it exhibited a lower overpotential of 330 mV and a current density of 10 mA cm^{-2} (Figures 7c, 7d). The primary cause of the higher activity was the synergistic interaction between N-GQDs and Co_3O_4. Figure 8 shows the SEM and TEM images of the prepared N-GQDs/Co_3O_4 [8].

Li et al. [41] synthesized $FeNi_3$@GQDs that exhibited exceptional electrocatalytic activity towards OER by showing a current density of 10 mA cm^{-2} with an overpotential of 238 mV, as well as a small Tafel slope of 48.7 mV dec^{-1}. The higher activity was attributed to the increased active areas, improved electrical conductivity, and strong synergistic coupling effect. From the DFT calculation, it was inferred that Fe was the most active site for accelerating water dissociation.

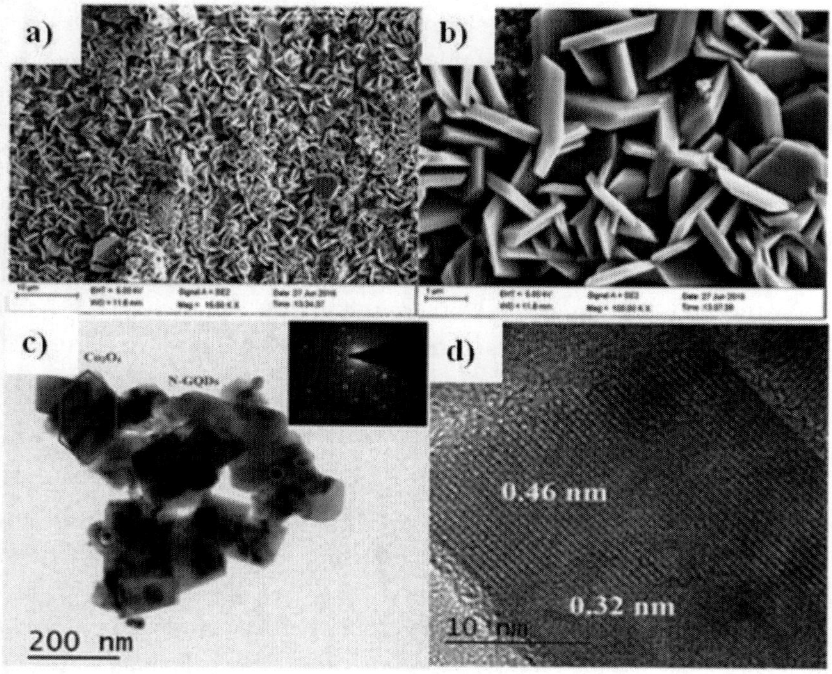

Figure 8. The SEM and TEM images of the prepared N-GQDs/Co_3O_4 [8]. Reprinted with permission from ref. [8]. Copyright 2017 Elsevier.

2.3. Oxygen Reduction Reaction (ORR)

Oxygen reduction reaction (ORR) typically involves the reduction of oxygen via a two-electron or four-electron pathway [42]. Two mechanisms are possible, in the two-electron pathway, H_2O_2 and HO^{2-} are formed in the acidic and alkaline media respectively, whereas in the four-electron pathway, H_2O and OH^- are formed. It finds application in metal-air batteries and fuel cells where the electrochemical reduction of oxygen is complemented by the generation of an electric potential. They involve high conversion efficiency with no pollution and hence have a large number of applications [42]. However, ORR taking place on the cathode of a fuel cell proceeds in the presence of a catalyst. In a typical proton exchange membrane fuel cell, oxidation of hydrogen on at anode gives electrons which aid in the reduction of oxygen on the cathode. Conventionally, Pt catalysts have been employed for the same. But these catalysts being highly expensive and susceptible to CO poisoning, are not suitable for commercial use [4, 9, 43]. Moreover, the sluggish kinetics of ORR in the presence of Pt based catalysts demanded a higher Pt loading. Hence, it was preferred to replace Pt with several earth-abundant metals and transition metal-free catalysts [4]. PtCu@GQDs capsules showed two-fold higher ORR activity and a 40 mV more positive onset potential as compared to conventional Pt catalyst [44]. Although GQDs have been shown to exhibit good activity in ORR, certain parameters like surface functionalization, doping, and number of active sites play a role in determining ORR efficiency [6]. Li and co-workers compared ORR activity of S-doped GQDs with varying amounts. Higher sulphur amounts led to greater electrocatalytic activity. It was also reported that with increasing thiophene S content, the electron transfer in the ORR process increased. However, no mechanism or explanation was given for the same [45].

Goswami et al. contrasted the ORR activity of N-doped, S-doped, and N/S co-doped GQDs in alkaline medium [6]. It was seen that doped GQDs had higher exchange current densities as compared to pristine GQDs. N-doped GQDS had the highest exchange current density, followed by N/S co-doped GQDs. Thus by CV measurements, it was inferred that N- doped GQDs were kinetically most favorable for ORR with a lower onset potential [6]. Guo and co-workers studied N doping on GQDs using computational tools [7]. It was reported that pyridinic N doping sites had a higher thermodynamic limiting potential for ORR, and hence exhibited higher electrocatalytic activity than graphitic N sites [7]. In another work, S/N co-doped GQDs decorated with TiO_2, supported on carbon, were employed for ORR. The catalyst was synthesized via a hydrothermal route, as shown in Figure 9. The synergistic interaction between S,N-GQD/TiO_2, and carbon, and the presence of sufficient oxygen deficient sites were the reasons proposed for enhanced ORR activity of the catalyst. The catalyst showed a Tafel slope of 61 $mVdec^{-1}$, methanol tolerance, and retention in current density up to 80% after 15 h [46]. DFT calculations were used to show that functionalization of graphitic N-doped GQDs with H_3O^+ and OH^- led to improved ORR activity [47].

Boron-doped GQDs anchored on graphene hydrogels prepared via a hydrothermal approach showed a unique trifunctional electrocatalytic activity for OER, HER, and ORR. The high stability of the catalyst was attributed to the 3D porous structure of graphene hydrogels, which prevented catalyst aggregation. The 3D structure also led to the exposure of active sites resulting in faster mass transport [20]. In another work, Wang et al. showed that B-doped GQDs on N-doped carbon nanotubes influenced the ORR activity positively, with an onset potential of 0.91 V. The catalyst was applied for the sensing of dissolved oxygen in seawater [48]. Another application of ORR was seen in microbial fuel cells [49]. B-doped GQDs were implanted on FeCo MOF supported on Ni foam via M-O-C bonding (BGQDs/MOF-15). The nanoflower-like structure, as seen in the SEM images (Figure 10a), led to greater active sites and low charge transfer resistance. The synthesized catalyst gave a power density of 703.55 mW m^{-2} when used as a cathode in microbial fuel cells [49]. In another work, boron and nitrogen co-doped GQDs supported on graphene (BN-GQD/G) were synthesized via a hydrothermal treatment [50]. The abundant sites, higher electrical conductivity, and surface area made the hybrid nanoplatelets excellent ORR catalysts having an onset potential of 15 mV. The uniform flake-like structure of the catalyst was confirmed based on the SEM (Figure 10b) and TEM (Figure 10c) images [50].

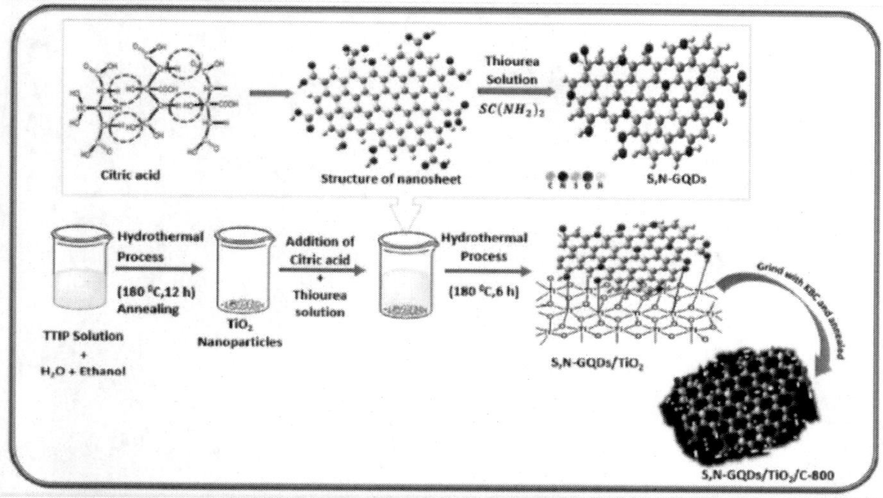

Figure 9. Synthesis of S,N-GQD/TiO_2/C-800 composite via hydrothermal approach [46]. Reprinted with permission from ref. [46]. Copyright 2021 Elsevier.

Zhang et al. synthesized GQD-Pt nanotube arrays (GQD–Pt NTAs) by a co-deposition method (Figure 11). The GQDs led to increased porosity of the catalyst, thus giving a high current density of 1.14 mA μg_{Pt}^{-1} at 0.90 V (vs. RHE). The greater availability of nanotube pores led to efficient transmission of active species thereby increasing the activity of the catalyst. The addition of GQDs prevented acid etching and also modified the surface of Pt electronically, preventing CO adsorption [51].

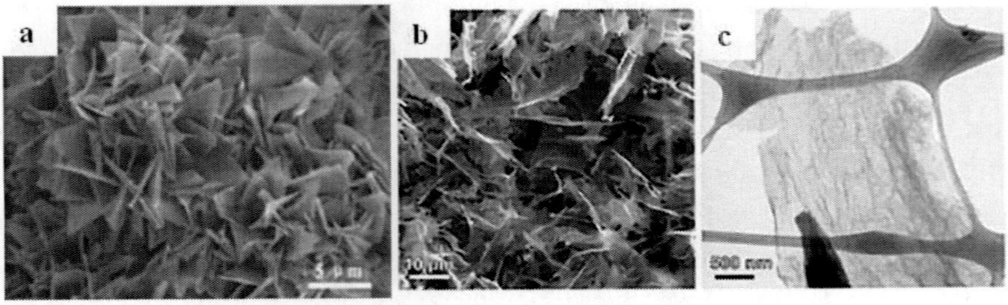

Figure 10. (a) SEM images for BGQDs/MOF-15 [49]. Reprinted with permission from ref. [49]. Copyright 2021 Elsevier. (b) SEM image of flake-like BN-GQD/G nanoplatelets, (c) TEM image of a typical individual BN-GQD/G nanoplatelet [50]. Reprinted with permission from ref. [50]. Copyright 2014 American Chemical Society.

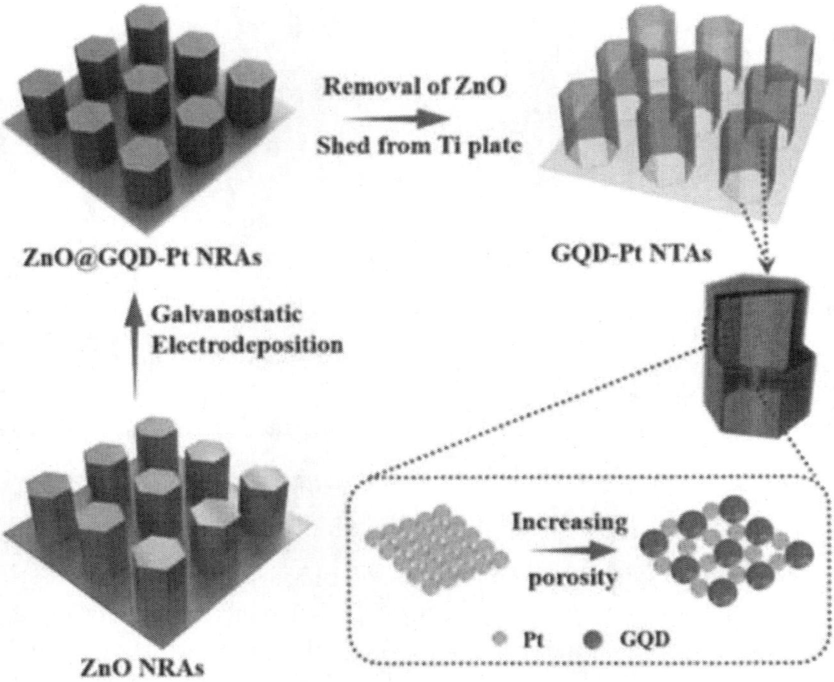

Figure 11. Schematic representation for the synthesis of GQD–Pt NTAs [51]. Reprinted with permission from ref. [51]. Copyright 2021 Royal Society of Chemistry.

Fan and co-workers illustrated that the addition of S to N-doped GQDs led to asymmetrical spin, charge density, and a change in the state of N, thereby causing a synergistic effect between N and S. The synergistic effect facilitated effective electron transfer in ORR. Figure 12a shows LSV curves for the synthesized composites, N and S-GQDs/rGO in comparison with the individual counterparts [52].

In another work, N-doped GQDs embedded into Co_3O_4 flakes were fabricated via a reduction process followed by annealing at 600°C [8]. Urea was used as a nitrogen source, and $NaBH_4$ was used as a reductant. From the TEM images, a uniform size distribution of the particles and the insertion of thin N-doped GQDs into Co_3O_4 flakes were seen. The catalyst showed bi-functional activity for both ORR and OER with lower onset potential, lower over-potential, and higher current density values (Figure 12b) [8]. Gao et al. utilized poly(3,4-ethylenedioxythiophene) (PEDOT) doped with GQDs, GQDs@PEDOT synthesized via electropolymerization, for ORR [53]. The addition of GQDs led to enhancements in surface area and current density of GQDs@PEDOT catalyst [53].

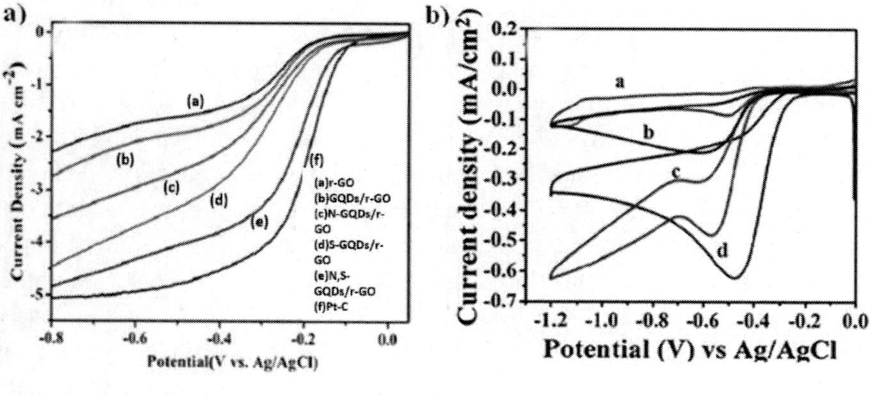

Figure 12. (a) LSV curves of rGO, GQDs/rGO, N-GQDs/rGO, S-GQDs/rGO, N-GQDs/rGO and Pt-C [52]. Reprinted with permission from ref. [52]. Copyright 2019 Elsevier. (b) CVs recorded for ORR using N-GQDs, bare GC, Co_3O_4, and N-GQDs/Co_3O_4 [8]. Reprinted with permission from ref. [8]. Copyright 2017 Elsevier.

Faraji and co-workers combined Pd_3Co with Fe-N doped GQDs for ORR. Although the catalyst showed 20% lower activity than Pt/C, a lower wt % (10 wt %) on the catalyst was required as compared to Pt/C (20 wt %). Pd_3Co/Fe-N-GQD retained 40% of its activity and was tolerant to methanol. The strong interface interaction between bimetallic Pd_3Co alloy nanoparticles and the support led to improved charge transfer resulting in increased ORR activity. Moreover, the porous nature of the support led to the easy adsorption of reactants on the catalyst [9].

Liu and co-workers demonstrated the synergistic effect between N-doped GQDs and $NiCo_2S_4$ in N-GQDs/ $NiCo_2S_4$. The enhancement in electrical conductivity obtained as a result of the synergistic effect facilitated the transport of reactants and products, thereby improving ORR kinetics [43]. A coral shaped catalyst formed by N doped GQDs and PtPd alloy, was also reported to give enhanced ORR activity. The metal support interaction, as well as the lattice strain and coordination effect in the PtPd alloy, favored easy adsorption of oxygen [54].

Conclusion

GQDs have been extensively researched for electrochemical OWS applications. Doped GQDs, as well as metal-GQDs based composites, have attracted the greatest attention for HER, OER, and ORR studies. This chapter has covered relevant GQDs based catalysts for OWS and metal-air batteries. The synergistic effect seen between GQDs and dopants, along with their charge transfer ability and numerous active sites, boosts the utility of GQDs. However, due to the difficulty in large scale synthesis, GQDs have still not been commercially exploited for hydrogen and oxygen production. Although GQDs are prominent for OWS applications, a greater insight into the mechanism involved at the molecular level is required. Various multi-doped GQDs can be further explored not only in OWS but also for other electrocatalytic and energy applications. Bifunctional GQDs based composites that give high efficiency in complete OWS require an extraordinarily stable electrocatalyst. This can be achieved by tuning the composition and morphology.

References

[1] Shaari, N.; Kamarudin, S. K.; Bahru, R. Carbon and Graphene Quantum Dots in Fuel Cell Application: An Overview. *Int. J. Energy Res.* 2021, *45* (2), 1396–1424. https://doi.org/10.1002/er.5889.

[2] Ali, A.; Shen, P. K. Recent Progress in Graphene-Based Nanostructured Electrocatalysts for Overall Water Splitting. *Electrochem. Energy Rev.* 2020, *3* (2), 370–394. https://doi.org/10.1007/s41918-020-00066-3.

[3] Riyajuddin, S.; Azmi, K.; Pahuja, M.; Kumar, S.; Maruyama, T.; Bera, C.; Ghosh, K. Super-Hydrophilic Hierarchical Ni-Foam-Graphene-Carbon Nanotubes-Ni2P-CuP2Nano-Architecture as Efficient Electrocatalyst for Overall Water Splitting. *ACS Nano* 2021, *15* (3), 5586–5599. https://doi.org/10.1021/acsnano.1c00647.

[4] Shao, M.; Chang, Q.; Dodelet, J. P.; Chenitz, R. Recent Advances in Electrocatalysts for Oxygen Reduction Reaction. *Chem. Rev.* 2016, *116* (6), 3594–3657. https://doi.org/10.1021/acs.chemrev.5b00462.

[5] Liu, W.; Li, M.; Jiang, G.; Li, G.; Zhu, J.; Xiao, M.; Zhu, Y.; Gao, R.; Yu, A.; Feng, M.; Chen, Z. Graphene Quantum Dots-Based Advanced Electrode Materials: Design, Synthesis and Their Applications in Electrochemical Energy Storage and Electrocatalysis. *Adv. Energy Mat.* 2020, *2001275*, 1–49. https://doi.org/10.1002/aenm.202001275.

[6] Goswami, M.; Mandal, S.; Pillai, V. K. Effect of Heteroatom Doping on the Electrocatalytic Properties of Graphene Quantum Dots for Oxygen Reduction Reaction Introduction. *Sci. Reps.* 2022, 1–16.

[7] Guo, Q.; Feng, J.; Chen, D.; Song, N.; Dong, H.; Yu, L.; Dong, L. Theoretical Insights into Enhanced Electrocatalytic Activity of Oxygen Reduction Reactions on N/S-Codoped Graphene Quantum Dots. *J. Phys. Chem. C* 2021, *125* (18), 9747–9755. https://doi.org/10.1021/acs.jpcc.1c01495.

[8] Muthurasu, A.; Sheen Mers, S. V.; Ganesh, V. Nitrogen Doped Graphene Quantum Dots (N-GQDs)/Co3O4 Composite Material as an Efficient Bi-Functional Electrocatalyst for Oxygen Evolution and Oxygen Reduction Reactions. *Int. J. Hydrogen Energy* 2018, *43* (9), 4726–4737. https://doi.org/10.1016/j.ijhydene.2017.11.157.

[9] Faraji, M.; Derakhshi, P.; Tahvildari, K.; Yousefian, Z. High Performance Fe and N-Codoped Graphene Quantum Dot Supported Pd3Co Catalyst with Synergistically Improved Oxygen Reduction Activity and Great Methanol Tolerance. *Solid State Sci.* 2018, *83*, 152–160. https://doi.org/10.1016/j.solidstatesciences.2018.07.012.

[10] Haque, E.; Kim, J.; Malgras, V.; Reddy, K. R.; Ward, A. C.; You, J.; Bando, Y.; Hossain, S. A. Recent Advances in Graphene Quantum Dots: Synthesis, Properties, and Applications. *Small Methods* 2018, *1800050*, 1–14. https://doi.org/10.1002/smtd.201800050.

[11] Rao, V. N.; Reddy, N. L.; Kumari, M. M.; Cheralathan, K. K.; Ravi, P.; Sathish, M.; Neppolian, B.; Raghava, K.; Shetti, N. P.; Prathap, P.; Aminabhavi, T. M.; Shankar, M. V. Sustainable Hydrogen Production for the Greener Environment by Quantum Dots-Based Efficient Photocatalysts: A Review. *J. Environ. Manage.* 2019, *248* (July), 109246. https://doi.org/10.1016/j.jenvman.2019.07.017.

[12] Horiuti, J.; Matsuda, A.; Enyo, M.; Kita, H. The Mechanism of the Hydrogen Evolution Reaction. In *Electrochemistry*; Elsevier, 1965; pp 750–779. https://doi.org/10.1016/B978-1-4831-9831-6.50063-1.

[13] Nemiwal, M.; Zhang, T. C.; Kumar, D. Graphene-Based Electrocatalysts: Hydrogen Evolution Reactions and Overall Water Splitting. *Int. J. Hydrogen Energy* 2021, *46* (41), 21401–21418. https://doi.org/10.1016/j.ijhydene.2021.04.008.

[14] Zeng, Z.; Chen, S.; Thatt, T.; Tan, Y.; Xiao, F. Graphene Quantum Dots (GQDs) and Its Derivatives for Multifarious Photocatalysis and Photoelectrocatalysis. *Catal. Today* 2018, *315* (September), 171–183. https://doi.org/10.1016/j.cattod.2018.01.005.

[15] Zhang, G.; Lan, Z. A.; Lin, L.; Lin, S.; Wang, X. Overall Water Splitting by Pt/g-C_3N_4 Photocatalysts without Using Sacrificial Agents. *Chem. Sci.* 2016, *7* (5), 3062–3066. https://doi.org/10.1039/c5sc04572j.

[16] Tsai, K.; Hsieh, P.; Lai, T.; Tsao, C.; Pan, H.; Lin, Y.; Hsu, Y. Nitrogen-Doped Graphene Quantum Dots for Remarkable Solar Hydrogen Production. *ACS App. Energy Mater.* 2020, *3*, 6, 5322–5332. https://doi.org/10.1021/acsaem.0c00335.

[17] Yan, M.; Hua, Y.; Zhu, F.; Sun, L.; Gu, W.; Shi, W. Constructing Nitrogen Doped Graphene Quantum Dots-$ZnNb_2O_6$/g-C_3N_4 Catalysts for Hydrogen Production under Visible Light. *Applied Catal. B, Environ.* 2017, *206*, 531-537. https://doi.org/10.1016/j.apcatb.2017.01.070.

[18] Raghavan, A.; Sarkar, S.; Nagappagari, L. R.; Bojja, S.; Venkatakrishnan, S. M.; Ghosh, S. Materials and Interfaces Decoration of Graphene Quantum Dots on TiO_2 Nanostructures: Photosensitizer and Cocatalyst Role for Enhanced Hydrogen Generation. *Ind. Eng. Chem. Res.* 2020, *59* (29), 13060−13068. doi: 10.1021/acs.iecr.0c01663.

[19] Gong, J.; Zhang, Z.; Zeng, Z.; Wang, W.; Kong, L.; Liu. J.; Chen, P. Graphene Quantum Dots Assisted Exfoliation of Atomically-Thin 2D Materials and as-Formed 0D / 2D van Der Waals Heterojunction for HER. *Carbon N. Y.* 2021, *184*, 554–561. https://doi.org/10.1016/j.carbon.2021.08.063.

[20] Tam, T. Van; Kang, S. G.; Kim, M. H.; Lee, S. G.; Hur, S. H.; Chung, J. S.; Choi, W. M. Novel Graphene Hydrogel/B-Doped Graphene Quantum Dots Composites as Trifunctional Electrocatalysts for Zn−Air Batteries and Overall Water Splitting. *Adv. Energy Mater.* 2019, *9* (26), 1–11. https://doi.org/10.1002/aenm.201900945.

[21] Tian, J.; Chen, J.; Liu, J.; Tian, Q.; Chen, P. Graphene Quantum Dot Engineered Nickel-Cobalt Phosphide as Highly Efficient Bifunctional Catalyst for Overall Water Splitting. *Nano Energy* 2018, *48*, 284–291. https://doi.org/10.1016/j.nanoen.2018.03.063.

[22] Zhao, M.; Zhang, J.; Xiao, H.; Hu, T.; Jia, J.; Wu, H. Facile: In Situ Synthesis of a Carbon Quantum Dot/Graphene Heterostructure as an Efficient Metal-Free Electrocatalyst for Overall Water Splitting. *Chem. Commun.* 2019, *55* (11), 1635–1638. https://doi.org/10.1039/c8cc09368g.

[23] Guo, B.; Yu, K.; Li, H.; Qi, R.; Zhang, Y.; Song, H.; Tang, Z.; Zhu, Z.; Chen, M. Coral-Shaped MoS_2 Decorated with Graphene Quantum Dots Performing as a Highly Active Electrocatalyst for Hydrogen Evolution Reaction. *ACS Appl. Mater. Interfaces* 2017, *9* (4), 3653–3660. https://doi.org/10.1021/acsami.6b14035.

[24] Lv, J. J.; Zhao, J.; Fang, H.; Jiang, L. P.; Li, L. L.; Ma, J.; Zhu, J. J. Incorporating Nitrogen-Doped Graphene Quantum Dots and Ni_3S_2 Nanosheets: A Synergistic

Electrocatalyst with Highly Enhanced Activity for Overall Water Splitting. *Small* 2017, *13* (24), 1–10. https://doi.org/10.1002/smll.201700264.

[25] Li, J.; Zhang, X.; Zhang, Z.; Li, Z.; Gao, M.; Wei, H.; Chu, H. Graphene-Quantum-Dots-Induced Facile Growth of Porous Molybdenum Doped Ni 3 S 2 Nanoflakes as Efficient Bifunctional Electrocatalyst for Overall Water Splitting. *Electrochim. Acta* 2019, *304*, 487–494. https://doi.org/10.1016/j.electacta.2019.03.023.

[26] Tsai, K.; Hsu, Y. Applied Catalysis B : Environmental Graphene Quantum Dots Mediated Charge Transfer of CdSe Nanocrystals for Enhancing Photoelectrochemical Hydrogen Production. *Applied Catal. B, Environ.* 2015, *164*, 271–278. https://doi.org/10.1016/j.apcatb.2014.09.034.

[27] Sim, Y.; Kim, S. J.; Janani, G.; Chae, Y.; Surendran, S.; Kim, H.; Yoo, S.; Seok, D. C.; Jung, Y. H.; Jeon, C.; Moon, J.; Sim, U. The Synergistic Effect of Nitrogen and Fluorine Co-Doping in Graphene Quantum Dot Catalysts for Full Water Splitting and Supercapacitor. *Appl. Surf. Sci.* 2020, *507*, 145157. https://doi.org/10.1016/j.apsusc.2019.145157.

[28] Wang, X.; Yuan, W.; Yu, Y.; Ming, C. Dynamically Self-Assembled Graphene Quantum Dots Mediated Synthesis of Truly Pristine Graphene Supported CoPNanoparticles as an Efficient Electrocatalyst for Hydrogen Evolution. *Chem Sus Chem,* 2017, *10*, 5, 1014-1021. https://doi.org/10.1002/cssc.201601761.

[29] Feng, Y.; Wang, X.; Huang, J.; Dong, P.; Ji, J.; Li, J.; Cao, L.; Feng, L.; Jin, P.; Wang, C. Decorating CoNi Layered Double Hydroxides Nanosheet Arrays with Fullerene Quantum Dot Anchored on Ni Foam for Efficient Electrocatalytic Water Splitting and Urea Electrolysis. *Chem. Eng. J.* 2020, *390*, 124525. https://doi.org/10.1016/j.cej.2020.124525.

[30] Xue, Y.; Xie, Y.; Xu, C.; He, H.; Jiang, Q.; Ying, G.; Huang, H. 0D/2D Heterojunction of Graphene Quantum Dots/MXene Nanosheets for Boosted Hydrogen Evolution Reaction. *Surfaces and Interfaces* 2022, *30* (March), 101907. https://doi.org/10.1016/j.surfin.2022.101907.

[31] Cao, C. T.; Kim, S. W.; Kim, H. J.; Purbia, R.; Kim, S. H.; Kim, D.; Choi, K. J.; Park, H.; Baik, J. M. N-Doped Graphene Quantum Dots as Charge-Transfer-Bridge at LaSrCoO/MoSe2 Heterointerfaces for Enhanced Water Splitting. *Nano Energy* 2022, *96* (March), 107117. https://doi.org/10.1016/j.nanoen.2022.107117.

[32] Gong, J.; Zhang, Z.; Xi, S.; Wang, W.; Lu, J.; Chen, P. Graphene Quantum Dot Enabled Interlayer Spacing and Electronic Structure Regulation of Single-Atom Doped MoS2 for Efficient Alkaline Hydrogen Evolution. *Chem. Eng. J.* 2023, *451* (P3), 138951. https://doi.org/10.1016/j.cej.2022.138951.

[33] Cheng, J.; Zhao, Z.; Zhang, B.; Xu, J.; Lin, C.; Li, X.; Gu, X. Stable and Active Low-Pt Electrocatalysts Using Defective Carbon Support Assembled from Graphene Quantum Dots. *J. Alloys Compd.* 2022, *900*, 163441. https://doi.org/10.1016/j.jallcom.2021.163441.

[34] Sharma, V.; Roondhe, B.; Saxena, S.; Shukla, A. Role of Functionalized Graphene Quantum Dots in Hydrogen Evolution Reaction: A Density Functional Theory Study. *Int. J. Hydrogen Energy* 2022, *47* (99), 41748–41758. https://doi.org/10.1016/j.ijhydene.2022.02.161.

[35] Ghosh, S. K.; Rahaman, H. *Noble Metal–Manganese Oxide Hybrid Nanocatalysts*; Elsevier Inc., 2019. https://doi.org/10.1016/B978-0-12-814134-2.00009-7.

[36] Pu, Z.; Liu, T.; Zhang, G.; Ranganathan, H.; Chen, Z.; Sun, S. Electrocatalytic Oxygen Evolution Reaction in Acidic Conditions: Recent Progress and Perspectives. *Chem Sus Chem* 2021, *14* (21), 4636–4657. https://doi.org/10.1002/cssc.202101461.

[37] Rossmeisl, J.; Qu, Z.-W.; Zhu, H.; Kroes, G.-J.; Nørskov, J. K. Electrolysis of Water on Oxide Surfaces. *J. Electroanal. Chem.* 2007, *607* (1–2), 83–89. https://doi.org/10.1016/j.jelechem.2006.11.008.

[38] Cirone, J.; Ahmed, S. R.; Wood, P. C.; Chen, A. Green Synthesis and Electrochemical Study of Cobalt/Graphene Quantum Dots for Efficient Water Splitting. *J. Phys. Chem. C* 2019, *123* (14), 9183–9191. https://doi.org/10.1021/acs.jpcc.9b00951.

[39] Alves, A. K.; Frantz, A. C. S.; Berutti, F. A. Microwave-Assisted Oleothermal Synthesis of Graphene-TiO_2 Quantum Dots for Photoelectrochemical Oxygen Evolution Reaction. *Flat Chem* 2018, *12* (December), 26–34. https://doi.org/10.1016/j.flatc.2018.12.001.

[40] Dong, Q.; Shuai, C.; Mo, Z.; Liu, Z.; Liu, G.; Wang, J.; Chen, Y.; Liu, W.; Liu, N.; Guo, R. Nitrogen-Doped Graphene Quantum Dots Anchored on NiFe Layered Double-Hydroxide Nanosheets Catalyze the Oxygen Evolution Reaction. *New J. Chem.* 2020, *44* (41), 17744–17752. https://doi.org/10.1039/d0nj03537h.

[41] Li, Z.; Xu, X.; Lu, X.; He, C.; Huang, J.; Sun, W.; Tian, L. Synergistic Coupling of FeNi3 Alloy with Graphene Carbon Dots for Advanced Oxygen Evolution Reaction Electrocatalysis. *J. Colloid Interface Sci.* 2022, *615*, 273–281. https://doi.org/10.1016/j.jcis.2022.01.088.

[42] Dai, L.; Xue, Y.; Qu, L.; Choi, H. J.; Baek, J. B. Metal-Free Catalysts for Oxygen Reduction Reaction. *Chem. Rev.* 2015, *115* (11), 4823–4892. https://doi.org/10.1021/cr5003563.

[43] Liu, W.; Ren, B.; Zhang, W.; Zhang, M.; Li, G.; Xiao, M.; Zhu, J.; Yu, A.; Ricardez-Sandoval, L.; Chen, Z. Defect-Enriched Nitrogen Doped–Graphene Quantum Dots Engineered NiCo2S4 Nanoarray as High-Efficiency Bifunctional Catalyst for Flexible Zn-Air Battery. *Small* 2019, *15* (44), 1–11. https://doi.org/10.1002/smll.201903610.

[44] Wang, L.; Hu, C.; Zhao, Y.; Hu, Y.; Zhao, F.; Chen, N.; Qu, L. A Dually Spontaneous Reduction and Assembly Strategy for Hybrid Capsules of Graphene Quantum Dots with Platinum-Copper Nanoparticles for Enhanced Oxygen Reduction Reaction. *Carbon N. Y.* 2014, *74*, 170–179. https://doi.org/10.1016/j.carbon.2014.03.021.

[45] Li, F.; Sun, L.; Luo, Y.; Li, M.; Xu, Y.; Hu, G.; Li, X.; Wang, L. Effect of Thiophene S on the Enhanced ORR Electrocatalytic Performance of Sulfur-Doped Graphene Quantum Dot/Reduced Graphene Oxide Nanocomposites. *RSC Adv.* 2018, *8* (35), 19635–19641. https://doi.org/10.1039/c8ra02040j.

[46] Mahato, D.; Kharwar, Y. P.; Ramanujam, K.; Haridoss, P.; Thomas, T. S, N Co-Doped Graphene Quantum Dots Decorated TiO2 and Supported with Carbon for

Oxygen Reduction Reaction Catalysis. *Int. J. Hydrogen Energy* 2021, *46* (41), 21549–21565. https://doi.org/10.1016/j.ijhydene.2021.04.013.

[47] Guo, Q.; Feng, J.; Liu, H.; Xia, C.; Dong, H.; Sun, Q.; Yu, L.; Dong, L. Effects of Hydronium and Hydroxide Ion/Group on Oxygen Reduction Reaction Electrocatalytic Activities of N-Doped Graphene Quantum Dots. *Mol. Catal.* 2022, *517* (October 2021), 112009. https://doi.org/10.1016/j.mcat.2021.112009.

[48] Wang, Y. X.; Rinawati, M.; Huang, W. H.; Cheng, Y. S.; Lin, P. H.; Chen, K. J.; Chang, L. Y.; Ho, K. C.; Su, W. N.; Yeh, M. H. Surface-Engineered N-Doped Carbon Nanotubes with B-Doped Graphene Quantum Dots: Strategies to Develop Highly-Efficient Noble Metal-Free Electrocatalyst for Online-Monitoring Dissolved Oxygen Biosensor. *Carbon N. Y.* 2022, *186*, 406–415. https://doi.org/10.1016/j.carbon.2021.10.027.

[49] Yan, Y.; Hou, Y.; Yu, Z.; Tu, L.; Qin, S.; Lan, D.; Chen, S.; Sun, J.; Wang, S. B-Doped Graphene Quantum Dots Implanted into Bimetallic Organic Framework as a Highly Active and Robust Cathodic Catalyst in the Microbial Fuel Cell. *Chemosphere* 2022, *286* (P3), 131908. https://doi.org/10.1016/j.chemosphere.2021.131908.

[50] Ihamäki, T. Susceptibility to Chronic Gastritis in First Degree Relatives of Gastric Carcinoma Patients. Effect of Histological Type of Carcinoma and Location of Tumour. *Ann. Clin. Res.* 1984, *16* (4), 183–187.

[51] Zhang, L.; Lu, P.; Luo, Y.; Zheng, J. Y.; Ma, W.; Ding, L.-X.; Wang, H. Graphene-Quantum-Dot-Composited Platinum Nanotube Arrays as a Dual Efficient Electrocatalyst for the Oxygen Reduction Reaction and Methanol Electro-Oxidation. *J. Mater. Chem. A* 2021, *9* (15), 9609–9615. https://doi.org/10.1039/D0TA12418D.

[52] Fan, T.; Zhang, G.; Jian, L.; Murtaza, I.; Meng, H.; Liu, Y.; Min, Y. Facile Synthesis of Defect-Rich Nitrogen and Sulfur Co-Doped Graphene Quantum Dots as Metal-Free Electrocatalyst for the Oxygen Reduction Reaction. *J. Alloys Compd.* 2019, *792*, 844–850. https://doi.org/10.1016/j.jallcom.2019.04.097.

[53] Gao, X.; Ma, J.; Li, Y.; Wei, H. Graphene Quantum Dots Doped PEDOT and Its Electrocatalytic Performance for Oxygen Reduction Reaction. *Int. J. Electrochem. Sci.* 2017, *12* (12), 11287–11297. https://doi.org/10.20964/2017.12.63.

[54] Liu, S.; Wang, W.; Hu, Y.; Tian, F.; Miao, X.; Liu, L.; Xu, Z. Hetero-Shaped Coral-like Catalysts through Metal-Support Interaction between Nitrogen-Doped Graphene Quantum Dots and PtPd Alloy for Oxygen Reduction Reaction. *Electrochim. Acta* 2020, *364*, 137314. https://doi.org/10.1016/j.electacta.2020.137314.

Chapter 6

Uses of Graphene Quantum Dots in Biological Applications

Behiye Senel[1,*]
and Mirsad Ibraimi[2]
[1]Department of Pharmaceutical Biotechnology, Anadolu University, Faculty of Pharmacy, Eskisehir, Türkiye
[2]TRPHARM, Istanbul, Türkiye

Abstract

In recent years, graphene quantum dots (GQDs) have attracted considerable research interest in the fields of medicine and biology. They appear to be suitable candidates for nanotheranostic (both diagnostic and therapeutic purposes) approaches because of their excellent biocompatibility and adjustable physicochemical properties. Graphene quantum dots have different properties according to their synthesis method and the starting material; thus, it is possible to design them as required for the specific use. Since they can be conjugated also with DNA/RNA, it allows us to use GQDs in applications such as gene silencing and gene regulation. This study will summarize the importance and innovations of GQDs synthesized from graphene in the fields of bioimaging, gene/drug delivery, along with the latest developments in these areas.

Keywords: graphene quantum dots, bioimaging, biosensing, gene/drug delivery

---

* Corresponding Author's Email: behiyek@anadolu.edu.tr.

In: Graphene Quantum Dots (GQDs)
Editor: Aleksey Kuznetsov
ISBN: 979-8-89113-638-0

Introduction

1. Quantum Dots (QDs)

QDs are theoretically defined as zero-dimensional, nanometer-sized, semiconductor particles, and most of their properties depend on their size, shape, and the material from which they are synthesized (Lee et al., 2018). QDs are used as potential markers in intracellular imaging due to their superior detection capabilities and color changes at different wavelengths in imaging (Lin et al., 2007). QDs can generally be synthesized from metals (such as gold), semiconductors (such as cadmium sulfide), or carbon-based materials (such as graphene) (Joshi et al., 2016). QDs consist of a semiconductor core and a shell surrounding it. Organic surfactants during synthesis form ligands on the surface.

Due to their nanosize, QDs exhibit a critical quantum confinement effect, resulting in a wide band gap and level gaps (Schmidt et al., 2004). Quantum dots are semiconductor nanoparticles that glow in different colours after being illuminated with a light source. When quantum dots are illuminated with UV light, some electrons get enough energy to leave the atoms, thus forming a conduction band where electrons can move freely through the material and conduct electricity and move around the nanoparticle. When these electrons return to the valence band, light can be emitted. The colour of this light changes according to the energy difference between the conduction band and the valence band. There is an inverse relationship between the nanoparticle size and the energy gap between the valence band and conduction band: as the nanoparticle size increases, this energy difference becomes smaller, and the light is red-shifted. On the contrary, when the QD size is reduced, the energy gap becomes large and a dark blue colour of QDs is observed. This property, which can also be seen in all other semiconductor materials, explains the QDs fluorescence.

It can be seen in the literature that QDs are also called "artificial atoms" (Zia-ur-Rehman et al., 2018). This is because a QD has bound and discrete electronic states similar to naturally occurring in atoms or molecules (Tisdale and Zhu, 2010). Moreover, thanks to the presence of active carboxylic acid esters around the QD core and shell, amine groups can bind to QD surface, enabling the transport of various aptamers and genetic materials due to electrostatic interactions (Samimi et al., 2017).

To summarize, due to QD unique particle size, broad absorption ability, size-adjustable emission, high photostability, and remarkable TEM contrast position, quantum dot-based therapeutic and diagnostic techniques can treat life-threatening diseases including cancer, renovascular and cardiovascular diseases, intestinal inflammatory diseases, and neurodegenerative diseases. emerging as a promising paradigm for combating these conditions (Tabish et al., 2018). QDs offer many advantages over organic dyes, including increased luminescence, broad absorption spectra, and tuneable narrow emission spectra (Wagner et al., 2019). Although they can be used as photodiodes due to their specific properties, their use in medicine and biology has been limited due to their potential toxicity (Nikazar et al., 2020). This limitation has led to the search for new alternatives and with the development of nanotechnology, QDs produced from graphene, or graphene quantum dots, have been designed as a more biocompatible source for biomedical applications.

2. Graphene Quantum Dots (GQDs)

The word "graphene" was first coined by A. K. Geim and K. S. Novoselov, and this material, which they isolated for the first time with the exfoliation method, brought them the Nobel Prize in 2004 (Xu and Lee, 2016). Carbon atoms are known, when bonded with four other carbon atoms, to be able to form three-dimensional lattices, giving rise to the substance called diamond, and when three carbon atoms combine with three other carbon atoms, they can form two-dimensional sheets called graphene. Graphene is a single layer of graphite which has stacked layers. After the discovery of graphene, 1-dimensional and then 0-dimensional (0D) graphene quantum dots with very high opto-electronic properties were developed (Jin et al., 2015). Graphene is a new material with a size range of 3-20 nm, molecule-like structure, relatively nontoxic, and easily processed compared to colloidal QDs (Şenel et al., 2019; Ying et al., 2022). Graphenes are synthesized in various geometric forms with round, elliptical, and polygonal shapes. The same structures made of carbon atoms bonded together in hexagonal shapes form carbon nanotubes, which are cylindrical compared to the flat graphene. Due to the quantum confinement effect, the difference in density, shape, and size of sp^2-hybridized fragments, the optical properties of GQDs are highly size-dependent and can be tuned by varying the energy band gap size. This allows GQDs to exhibit unusual optical and electrical properties unique to them (Mueller et al., 2010). For example, GQDs are known that emit blue, green, yellow, and red ultraviolet colours according to the band gap value (Ke at el., 2015; Al Ghamdi and Al-Ghamdi,

2023). In other words, thanks to the quantum confinement effect, the colour of the radiation GQDs emit changes as the size of the quantum dots changes.

While GQDs show similar benefits to conventional QDs, they are less toxic, mainly because they consist of carbon and do not contain heavy metals, and are highly soluble in water due to the different functional groups present on their surface, which makes them suitable for diagnostic and therapeutic use for diseases such as cancer. In the study reported in the literature, GQD-based anticancer drug carrier was prepared and reported to show the drug delivery and release by emitting fluorescent signals at different levels (Ding et al., 2017). Moreover, the barrier for crossing the BBB (Blood-Brain-Barrier), which causes the low efficacy of many drugs, does not apply to GQDs (Tabish et al., 2018). GQDs are advantageous species in drug delivery compared to other nanoparticles in terms of their small size and lipid-soluble properties. They have the ability to be easily targeted to the brain with transferrin ligand or glucose transport protein (Perini et al., 2020; Demirdöğen, 2022).

2.1. Synthesis Methods

GQDs, like other biotechnological systems, are synthesized in two ways, either as a top-down or bottom-up approach. In top-down methods, two-dimensional (2D) graphene or graphene oxide (GO) sheets, carbon fibres, carbon nanotubes (CNTs), or graphite are broken down to form zero-dimensional (0D) GQDs, while molecular precursors are synthesized in bottom-up processes, with small stepwise reactions (Mansuriya and Altıntaş, 2020). In the top-down approach, zero-dimensional GQDs are prepared by electron beam lithography, arc-discharge methods, using plasma reactors, electrochemical oxidation, solvothermal process, microwave-assisted hydrothermal cutting/reaction, microfluidization and electrochemical/liquid exfoliation methods, and this synthesis is performed using carbon source materials such as graphene oxide, graphene nanoribbons, graphene sheet, graphite, citric acid, coal, or carbon fibre (Shaari et al., 2020). In the bottom-up approach, the precursors that will form the GQDs are brought together into nanocrystals with various physical and chemical approaches (Wang et al., 2015).

2.2. Characterisation of GQDs

Similar characterization studies are carried out for GQDs as for other QDs. In characterization studies, properties such as particle size, optical properties, surface chemistry, excitation and emission spectra, quantum yield, and brightness are examined. For example, the particle size of QDs plays a crucial role in determining their optical properties. For this, DLS (dynamic light scattering) is the basic technique (Drbohlavova et al., 2009). However, considering the structure of graphene, there is a need for confirmation of DLS results by electron microscopy methods such as SEM (scanning electron microscopy), TEM (transmission electron microscopy) or AFM (atomic force microscopy), or photoluminescence, or Raman spectroscopy. In the examination of the surface chemistry of GQDs, which is another important component in their characterization, such techniques as photoelectron spectroscopy, nuclear magnetic resonance (NMR) spectroscopy, Rutherford backscattering, and analytical ultracentrifugation can be used (Lees et al., 2008). Förster resonance energy transfer (FRET), thermogravimetric analysis (TGA), UV-vis spectrophotometry, and photoluminescence spectrometry are also the leading tools in determining optical properties, excitation and emission spectra, quantum efficiency, and brightness (Drbohlavova et al., 2009).

Due to the excellent optical properties of GQDs, it is possible to examinate tissue cell transitions in vitro by confocal microscopy or fluorescent microscopy, as well as to analyse biodistribution in experimental animals *in vivo* by imaging systems (Texier and Josser, 2009; Szymanski et al., 2013).

3. Biological Applications

3.1. Use of GQDs as Biosensors

Living organisms communicate with their environment with the help of physical and chemical biological sensors in their bodies. The living body is itself a biological sensor, because when the balance of organic and inorganic substances in a living thing is disturbed or a foreign substance enters the body, various diseases are seen in the living organism (Saffioti et al., 2020). In other words, the body reacts like a sensor according to the deviations of the substances present in the body from the balanced amount. However, in order to obtain clear information about the amount of any substance in the body, very sensitive biological sensors that convert biological signals into electrochemical, electrical, optical, gravimetric, or acoustic signals are needed (Naresh and Lee, 2021).

Biosensors are used in many areas that concern living things. In order for a sensor to be called a biosensor, it must be given a biofunction. The biofunctionalization process determines the working area of the sensor because the factor that makes the sensor selective against biological substances consists of biological substances such as enzymes, proteins and DNA attached to the sensor surface. In this case, the material attached to the sensor surface should be selected in accordance with the working area (Bhala et al., 2016).

Diagnosing diseases at an early stage is essential to increase survival rates. Today, various imaging techniques such as ultrasound imaging, MRI (Magnetic Resonance Imaging), CT (Computerized Tomography), PET (Positron Emission Tomography) for early diagnosis are traditional imaging tools used for this purpose. Unfortunately, these tools have some limitations, for example, poor resolution, poor cell penetration, problems with real-time imaging, body exposure to high radiation, good expertise in acquisition procedures, or patient compliance (Forte et al., 2020). For this reason, the search for alternative imaging methods has been growing recently. To the date, nanowires, nanoribbons, carbon nanotubes, graphene, magnetic beads, and quantum dots have been among the major nanoparticular systems used to further improve the sensitivity, selectivity, and reproducibility of electrochemical immune sensors (Zumpano et al., 2021). Among these materials, QDs such as carbon dots and GQDs, which are relatively easy to functionalize and synthesize, have good biocompatibility, electrocatalytic activity, controllable size, good signal amplification and multiplexed sensing ability, chemical stability, water solubility, and inertness (Mansuriya et al., 2019). GQDs have gained popularity due to their unique properties such as resistance to flashing and light bleaching (Sun et al., 2013; Faridbod and Sanati, 2016; Raj and Jhon, 2019).

Recently, the excellent biocompatibility of GQDs, their stable photoluminescence (PL), resistance to light exposure, low *in vitro* and *in vivo* cytotoxicity, physiological stability, easy passage even through barriers such as BBB, non-functional surface, or adaptive fluorescent properties facilitated their use in this field as well. Conventional imaging products, such as fluorescent dyes are easily bleached by light, limiting their application due to their chemical instability. Particles with novel fluorescent properties, such as quantum dots, are stable products with fluorescent properties, but their cytotoxicity is their disadvantage. For this reason, researchers have been recently trying to eliminate these disadvantages with photostable graphene quantum dots (Nikazar et al., 2020). Therefore, GQDs appear as more reliable and useful species and are more widely used in biomedical imaging and labelling probes (Cai et al., 2015). GQDs are used as active sensor elements in high resolution cellular imaging, where the fluorescence properties of quantum dots change after reacting with the analyte or are used in passive label probes where selective receptor molecules such as antibodies are conjugated to its surface (Mansuriya and Altıntaş, 2020).

Cui et al. (Cui et al., 2019) prepared photoluminescent GQDs for fast, efficient, and sensitive separation and detection of circulating tumour cells (CTCs), and they stated that these nanoparticles are excellent both for capturing tumour cells and for bioimaging in the relevant region. In this study for the detection of CTCs, GQDs were used to prepare a "turn-on" magnetic fluorescent biosensor based on Fe_3O_4 and molybdenum disulphide (MoS_2) nanolayers and a detection limit of 1.19 nM was achieved in the linear range between 2 and 64 nM.

Among the GQDs, nitrogen-doped GQDs (N-dopped GQDs) are also noteworthy. Nitrogen atoms attached to GQDs can give rise to unexpected new properties, especially by changing the electronic properties and increasing the number of anchor sites for the adsorption of metal ions. For example, the Eudragit polymer coated nitrogen-doped GQDs prepared by Şenel et al. could be easily observed in cell culture with a confocal microscope. Thus, it has been determined that drug-loaded nanoparticles can be easily monitored inside the cell (Şenel et al., 2019). In another study, in which an electrochemical immune sensor was prepared with nitrogen-doped graphene quantum dots with the green hydrothermal method, the quantitative detection of carcino-embryonic antigen (CEA) could be performed at an ultra-sensitive level (Yang et al., 2017). Ganganboina et al. (Ganganboina et al., 2019), for the sensitive and selective detection of CEA, developed a label-free impedimetric immune sensor based on nitrogen- and thiol-doped GQDs (N,S–GQDs) and gold-embedded polyaniline (Au–PANI) nanowires. They reported that the GQD-based label-free nanowires increased electron transfer in a wide linear range between 0.5 and 1000 ng mL^{-1} with a detection limit (LOD) of 0.01 ng mL^{-1} and that CEA could be determined with excellent recovery of (96.0 ± 2.6)–$(103 \pm 3.8)\%$.

Another study by Kalkal et al. (Kalkal et al., 2020) used for ultra-sensitive detection of small cell lung cancer biomarkers biofunctionalized N-doped GQDs as energy donors and AuNPs as energy acceptors. The LOD value of this biosensor developed in the study was 0.09 pg/mL, the detection range was 0.1 pg mL^{-1} to 1000 ng mL^{-1}, and the response time was 16 minutes.

GQDs have been used as biosensors not only for bioimaging for diseases such as cancer, but also for the diagnosis of important diseases such as cardiovascular diseases, infectious agents, or for the detection of various other biomarkers in the body. For example, in the study reported in the literature (Mollarasouli et al., 2018), a new label-free electrochemical immunosensor was developed for the selective and sensitive detection of triose receptor kinase AXL, which is an important biomarker in heart attack, in serum. This disposable immunosensing platform was prepared by immobilization of specific anti-AXL antibody on screen-printed carbon electrodes (SPCEs) modified with amine functionalized graphene quantum dots (fGQDs). The prepared immunosensor showed a wider range of linearity than other electrochemical immunosensors reported so far, and a detection limit (0.5 pg mL^{-1}) in serum more than a hundred thousand times lower than the cut-off value (71 ng mL^{-1}) for the diagnosis of heart failure. In another study, an electrochemical biosensor based on a glassy carbon electrode coated with graphene quantum dots was developed for the detection of C-reactive Protein (CRP), an acute phase reactant produced by liver and fat cells, in serum, which increases in blood during inflammatory reactions. It was stated that with the developed biosensor, a low detection limit of 176 pM of CRP in serum could be determined (Bing and Wang, 2018).

In the study by Yan et al. in 2018 (Yan et al., 2018), a label-free electrochemical immunosensor was designed to quantitatively detect Hepatitis B surface antigen (HBsAg). In this study, the biosensor composed of nitrogen-doped graphene quantum dots (N-GQDs) and surfactant-free AuPdCu ternary nanoparticles with good conductivity and excellent catalytic properties for hydrogen peroxide (H_2O_2) reduction was made, providing excellent selectivity, reproducibility, and stability. With this developed biosensor, HBsAg was detected with high sensitivity and stability with a detection rate of 3.3 fg/mL (signal-to-noise ratio 3).

In another study by Roushani et al. (Roushani et al., 2019), GQDs were used for the first time for ultrasensitive recognition of Human Chorionic Gonadotropin (hCG) in serum samples. In this method, an immunosensor was produced based on screen-printed carbon electrode (SPCE) modified with GQDs-N-S/Au nanoparticles (NPs), which has low cost, shortens the processing time, and can make sensitive measurements without the need for any other substrate.

In 2019, a new, rapid, ultra-sensitive, and highly specific, label-free immunosensor approach was developed for the effective diagnosis of the infectious agent Yersinia enterocolitis. This sensor was prepared by enriching the gold electrode surface with GQDs. With the prepared sensor, a wide concentration range in complex media such as milk and human serum could be determined, and a very low detection limits of 5 cfu mL^{-1} and 30 cfu mL^{-1}, respectively, were found. The authors described the prepared GQD sensor as an attractive analytical tool that could pave the way for the identification of any pathogenic bacteria in clinical samples and food samples (Savaş and Altıntaş, 2019).

Moreover, in the different study by Niu et al. (Niu et al., 2018), heavy metal detection such as Pb^{2+}, which is important for both the environment and human health, was successfully detected with a sensing technology based on graphene quantum dots and gold nanoparticles. In this way, the researchers claim that with the results they obtained, an opportunity for rapid examination may arise not only for heavy metals such as Pb, but also for antigens or DNAs in chemical or biological studies.

3.2. Drug Delivery Applications of GQDs

The nanoparticular systems used for drug delivery system can be classified as polymeric (polymersome, dendrimer, polymeric micelles, nanospheres etc.), lipidic (liposome, emulsion, solid lipid nanoparticles, nanostructured lipid carriers etc.), and inorganic systems (silica, magnetic nanoparticles, quantum dots etc.) (Crintea et al., 2022). All these systems developed to date have their own advantages and disadvantages. For example, liposomes are biodegradable and obtained from biocompatible materials, have high encapsulation abilities, the ability to carry polar/nonpolar molecules, however, they have low stability and the disadvantages of creating an immune reaction (Nsairat et al., 2022). Gold or other inorganic nanoparticles are biocompatible, have large surface area and sustained release, but are not biodegradable, and their toxicity may be high (Yusuf et al., 2023). In general, polymeric nanoparticles show excellent stability, allow controlled synthesis and conjugation of many bioactive agents, but have disadvantages such as agglomeration and nanotoxicity with non-biodegradable polymers (Ratemi et al., 2016).

Among the most important features desired in nanoparticulate systems used as a drug delivery system are that they are biocompatible, that the drug can be delivered to the target area in sufficient quantity, and that this transport system can then be excreted by the kidneys at the appropriate time. However, these issues are not obstacles that can be overcome immediately, therefore research is still continuing rapidly to develop the best delivery system and new alternative ways are sought (Yusuf et al., 2023). For this reason, quantum dots have started to take their place among drug delivery systems. GQDs have recently started to attract attention in drug delivery systems, in addition to their use as biomarkers in areas such as bioimaging, especially due to their unique properties. In the studies reported, GQDs appear as promising materials for drug delivery systems due to their smaller size properties, large surface areas, excellent fluorescence capabilities, biocompatibility, and photostability. More importantly, GQDs can provide π-π interactions with drug molecules and can be functionalized, have some special characteristics such as an oxygen-rich surface, and can be represented as a single atomic layer with a small lateral size which makes them an alternative for the drug delivery process and enhances physiological stability (Hui et al., 2016.; Biswas et al., 2021).

The GQD fluorescent features allows traceability of drugs in cancer cells. For this reason, GQDs gave been widely used in drug transport in various diseases (Cheng et al., 2017). For example, in the study by Tian et al. (Tian et al., 2017), nanoparticles prepared with zeolite imidazolate framework (ZIF-8) and GQD as an efficient drug carrier were loaded with doxorubicin and it was stated that the loaded DOX (Doxorubicine) both exhibit a very good death on cancer cells with the π-π interactions with drugs and allow the drug release of doxorubicin to be monitored thanks to the GQDs. In another study, Lawson (2-hydroxy-1,4-naphthoquinone) was used as a new reducing agent for the synthesis of quantum dots conjugated with albumin. Gemitabine was loaded into the prepared GQD nanoparticles and both death and imaging of pancreatic cancer cells was observed (Nigam et al., 2014).

In one more study, it was determined that GQDs prepared by photo-Fenton reactions of graphene oxide loaded with cisplatin facilitate the drug transition into the cell and nucleus and increase the interaction of cisplatin with DNA (Sui et al., 2017). Also, a chemotherapy agent, methotrexate, was loaded on nitrogen-doped GQDs synthesized by Khodadadei et al. (Khodadadei et al., 2017). They prepared GQDs by a hydrothermal method by pyrolysis of citric acid as carbon source and urea as nitrogen source. They showed using FT-IR and UV spectroscopies that nanoparticles had strong π-π stacking and showed more activity than MTX (Methotrexate) alone.

Among the most important features expected from the nanoparticular systems used for drug delivery systems, it is expected to be able to target the drug to required organs, to have long retention in blood circulation, to be able to cross the biological barrier, to accumulate around the tumor, to assist cell entry, and to assist the release of the drug in the acidic tumor environment, which also causes the degradation of many drugs (Zhao et al., 2020). There are studies that report the feature of targeting for the GQD complexes prepared during the studies. For example, it was determined that folic acid, which is attached to a GQD nanoparticle loaded with doxurubicin, allows more selective input of the nanoparticles into the cells and increases accumulation on the target and the effect of doxorubicin (Wang et al., 2014). In another study by Wei et al. (Wei et al., 2018), GQDs prepared by chemical oxidation of dried CX-7 carbon black with HNO_3 were first loaded with cisplatin and then were grafted at PEG (Polyethylene Glycol), which would allow this complex to remain in the circulation for a long time and escape from the circulating reticulo endothelial system (RES) cells. This complex, which was obtained at a size of approximately 5.72 nm, was found to have a good therapeutic effect in oral squamous cell cancers in normoxic (normal oxygen concentration) and hypoxic (low levels of oxygen concentration) conditions when applied to mouse xenograft models. Dong et al. (Dong et al., 2017) loaded doxorubicin to another GQD complex which they prepared as conjugate with arginine-glycine-aspartic acid (RGD) peptide for both targeting and imaging, and the complex released the drug in a pH-sensitive manner due to the hydrogen bonds between GQD and doxorubicin. They suggested that compared to free doxorubicin, the complex accumulates much more in U251glioma cells within 16 hours and creates a more cytotoxic effect on these cells, so that treatment can be performed without the need for extra dosing. In another study, carbon black was used as a precursor in the preparation of GQDs, and oxidation and ball milling processes were used as the synthesis method. This system, in which B-Lapachone is loaded via the π-π bond interactions, was tested for different cancer cells such as LO2, HeLa, PC-12 and MCF-7. A higher cytotoxic effect was determined on system cells with high transfection characteristics compared to the free drug (Prabhakar et al., 2022).

3.3. Gene Delivery Applications of GQDs

Although the majority of GQD applications are focused on bioimaging, studies of tumor-targeted drug release have recently increased (Dong et al. 2018; Qi et al., 2021; Yu et al., 2021). Developing technology has led to the regulation of new treatments in the field of health, and gene therapies, in which genes are used in the treatment of diseases, have begun to gain prominence (Moghadam et al., 2023). The main purpose of the gene therapy is to regulate gene expression by either adding new gene or removing existing damaged gene and inserting new therapeutic gene to induce gene expression in order to treat disease. In such a treatment, nanoparticles are used to transport DNA/RNA to cells. However, the large size of these macromolecules makes them difficult to enter into the cell and nucleus. GQDs are excellent candidates for gene carriers due to their small size, ability to cross the blood-brain barrier, low toxicity, good stability, and luminescence properties that allow easy monitoring of gene/drug release (Ghafary et al., 2017).

After it was understood that GQDs could be an important transport system for DNA or RNA, studies were started on this subject, and it is predicted that these studies will increase even more. For example, in a study in which plasmid DNA was conjugated with MPG-2H1 chimeric peptide and GQDs via non-covalent interactions, using the green and red emissions of GQDs, this complex could be easily observed in HEK293T cells under a confocal microscope. Researchers reported that this 280 nm complex prepared with GQDs had low cytotoxicity and high transfection properties (Ghafary et al., 2017).

In another study, in which glycine-proline-glutamate as a neuroprotective peptide was conjugated with GQDs, the complex was targeted to the central nervous system, and as a result of the study it was shown that the prepared complex could prevent amyloid-β (Aβ) aggregation and reduce the inflammatory response, protect the synapses, and promote neurogenesis (Xiao et al., 2016).

In one more study, the biopolymer was functionalized by adding polyethyleneimine (PEI) to greatly improve the bioavailability of GQDs and even conjugated with folic acid to further increase the transfection efficiency. In this way, the aim was both to increase biocompatibility and to help keeping the complex highly stable under the acidic tumour area. Then, doxorubicin active substance and green fluorescent protein (GFP) were added to the formed complex. In this study and in other studies, the genetic material is generally loaded onto the complex via electrostatic interaction. In xenograft mice models, it exhibited a much higher activity on cancer cells than the free drug, with high biocompatibility and transfection efficiency in denier performed on colon cancer (Lo et al., 2020).

RNA interference is a biological mechanism by which double-stranded RNA (dsRNA) targets mRNA for degradation, causing gene silencing. First discovered in plants, this mechanism is a major innovation in the universal therapeutic treatment of diseases and revolutionizing the way researchers study gene function. Small interfering RNA (siRNA) and micro-RNA (miRNA) are double-stranded RNA molecules that have 20-25 base pairs in length and are involved in the RNA interference pathway. They are tasked with interfering with the expression of specific genes with complementary nucleotide sequences, by cleaving mRNA after transcription and preventing translation. Although siRNA is similar in principle to miRNA, it is a molecule with high target specificity due to its ability to intervene before translation and to have 100% complementarity. In the study by Şenel et al. (Şenel et al., 2019), nitrogen doped GQDs were prepared by the hydrothermal reaction between citric acid and p-aminophenol, and because these complexes have negative zeta potential, the complex was made positive with Eudragit polymer and

EPHA2 siRNA was added. The complex effects on lung and breast cancer cells were evaluated. As a result, the prepared complexes were internalized very well into the cell and high activity was detected with both DNA and mRNA degradation. In another study by Liu et al., EEGFP mRNA was prepared with citric acid using the typical bottom-up method and loaded into GQDs functionalized with PEI. As a result of the study, it was determined that even at very low doses, there was an internalisation to Huh-7 hepatocarcinoma cells (Liu et al., 2021).

3.4. Photodynamic Therapy (PDT)

Photodynamic therapy is a non-invasive treatment used in medicine. Photodynamic therapy is a treatment to treat skin, head and neck, tongue, and breast. Photodynamic therapyis a method applied using light and photo-stimulating special drugs to kill cancer cells (Tan et al., 2022). When photosensitizers are exposed to light of a certain wavelength, they undergo photoactivation and cause the generation of reactive oxygen species (ROS), which can cause cell damage and cell death. These reactive oxygen species (ROS) also cause the destruction of cancer cells. In the first stage of PDT, a photosensitizing agent is injected into the bloodstream. This drug is absorbed by cells throughout the body but stays longer in cancer cells. During this period, the patient is asked not to go out into the sunlight and stay in the dark. About 24 to 72 hours after injection, the tumour is exposed to light when most of the photosensitizer leaves the normal cells and remains in the cancer cells. The photosensitizer in the tumour absorbs the light and destroys the tumours by producing an active form of oxygen that destroys the nearby cancer cells (Zhou et al., 2016). GQDs have also been used in PDT therapy because of their photosensitizing effects, as described above. GQDs were prepared by Markovic et al. (Markovic et al., 2012) for the treatment of human glioma by excitation with blue light, which can produce ROS species when stimulated with blue light. The aim of the study was to accelerate the process of photosensitized cells to undergo apoptosis and autophagy, and to increase the clearance of oxidatively damaged organelles. It was observed in the analyses that the prepared GQDs caused an increase in both early and late apoptotic cell populations by inducing oxidative stress depending on the concentration and irradiation exposure time. In yet another study, Wang and colleagues (Wang et al., 2019) conducted very important research. In their study, boron-nitrogen-doped GQDs complex were irradiated by an external NIR source. As a result of this irradiation, GQDs showed the ability to absorb NIR radiation and converted it into heat, causing the death of cancer cells. In the study, a completely suppressive therapeutic effect was observed with photothermal effect in glioma xenograft mice. In the study in which GQDs were administered intravenously, it was observed that tumour growth was completely stopped at the end of 14 days.

In the study by Ahirwar et al. (Ahirwar et al., 2020), GQDs/GOQDs were synthesized by electrochemical exfoliation of graphite rods, while using a combination of citric acid, NaOH, and H_2O_2 at different ratios. In this study, 365 nm UV radiation was used. The effects of the GQD complex on skin cancer (B16F10) and breast cancer (MCF-7) cells were evaluated and it was determined that the prepared complexes showed easy cellular uptake and distribution with low cytotoxicity, but the penetration depth of UV irradiation was low. It was found that over 90% cell death occurred even in a short time as 5 minutes.

As a result, the number of studies is limited because the use of GQDs in the treatment of PDT is new. However, only one study showed a result that did not show effectiveness (Chen et al., 2020). For this reason, it is thought that whether GQDs can really be used in this therapy will be revealed with more research to be done in the coming years.

Conclusions and Future Perspectives

The discovery of the extraordinary and versatile properties of nano pharmaceuticals over time leads to the development of new drug delivery systems. In this way, for many diseases such as cancer nano pharmaceuticals have and will continue to have revolutionary effects on both diagnosis and treatment. Nano pharmaceuticals have taken their place as important materials not only in drug/gene delivery in disease treatment, but also in other medical fields such as diagnostic imaging of diseases and tissue engineering. Nanocrystals, liposomes and lipid nanoparticles, PEGylated polymeric nano-drugs, other polymers, protein-based nanoparticles and metal-based nanoparticles draw attention among these nano pharmaceuticals both already in the market and still studied today. The most important or desired features in the development of new formulations are the characteristics of these systems to increase the bioavailability of the molecules they carry, improve their distribution, regulate their pharmacokinetic and pharmacodynamic properties, ensure that they remain in the desired area for the desired time, act for the desired time, not cause any immunological response, not cause toxic effects, and of course to have low cost. To date, numerous formulations have been developed in many research and development laboratories, but most of them fail due to disadvantages such as various enzyme degradations or inability to adequately help the bioavailability of drugs, inability to reach the target site, immunogenicity, solubility problems, rapid clearance, and large molecular size.

One of the most important goals in pharmaceutical research is the synthesis or discovery of new chemical entities with minimum side effects and maximum clinical benefit. One of materials that emerged among these discoveries are graphene quantum dots. GQDs are carbon-derived nanomaterials with a quantum confinement effect and the edge (armchair or zigzag) effect, consisting of zero-dimensional very thin sheets of graphene. GQDs have attracted great attention because they show low toxicity, high biocompatibility, high fluorescent activity, strong chemical inertness, and excellent photostability, so that they can be a DDS (Drug Delivery Systems) to be used in drug/gene transport and they also provide the possibility of simultaneous treatment, detection, and imaging. In particular, these species, which can respond to certain internal or external physical or chemical stimuli such as light, temperature, pH, are indeed attracting medical attention. For example, considering the fact that cancer cells have a more acidic microenvironment compared to normal cells, the advantages of GQDs are undeniable, such as releasing the drug that they carry, especially pH sensitive, providing synergistic affect with chemotherapeutic treatments with the use of their photophysical properties, providing both treatment and imaging, and targeting by conjugation with antibodies/proteins. The only disadvantages of GQDs are their cytotoxic effects, which are still controversial. In the last study published in 2022, it was seen that there are studies only for QDs in clinical research Phase I and Phase II stages. In the article in question, it was mentioned that the only QDs investigated for bioimaging in studies were those coded NCT04138342, NCT04390490, NCT03895437, NCT03794973, NCT03794960, NCT04590872 (Abdellatif et al., 2022). Only on the website of clinicaltrilas.gov.tr, in the study with the code NCT05841862, there is a QDs device study started in 2023 to improve visual function in adults with Advanced Retinitis Pigmentosa. Apart from this, for GQDs, there is currently no case that has passed to the clinical phase study on drug/gene transport. Again, GQDs are thought to have the security concerns. Considering the studies, it is thought that their cytotoxicity is due to their dimensions, electrical charge, or synthesis method. For this reason, as in all nano pharmaceuticals, toxicity in GQDs seems to depend on the synthesis and starting material. Thus, when the use of green resources or of hydrothermal production processes is ensured in synthesis, its use as a new generation drug carrier and imaging tools will become widespread. More extensive research and development of practical applications would be good for the further development of GQDs field.

References

Abdellatif, A. A. H.; Younis, M. A.; Alsharidah, M. et al. Biomedical Applications of Quantum Dots: Overview, Challenges, and Clinical Potential. *Int. J. Nanomed.* 2022, *17*, 1951–1970.

Ahirwar, S.; Mallick, S.; Bahadur, D. Photodynamic therapy using graphene quantum dot derivatives. *J. Solid State Chem.* 2020, *282*, 121107,1-9.

Al Ghamdi H. S.; Al-Ghamdi, A. A. Opening band gap of multi-color graphene quantum dots using D-fructose as a green precursor. *Alex. Eng. J.* 2023, *79*, 155-163.

Bhalla, N.; Jolly, P.; Formisano, N. et al. Introduction to biosensors. *Essays Biochem.* 2016, *60*(1), 1-8.

Bing, X.; Wang, G. Label Free C-reactive Protein Detection Based on An Electrochemical Sensor for Clinical Application. *Int. J. Electrochem. Sci.* 2017, *12*, 6304-6314.

Biswas, M. C.; Islam, M. T.; Nandy, P. K. et al. Graphene Quantum Dots (GQDs) for Bioimaging and Drug Delivery Applications: A Review. *ACS Mat. Lett.* 2021, *3*(6), 889-911.

Cai, Z.; Li, F.; Wu, P. et al. Synthesis of Nitrogen-Doped Graphene Quantum Dots at Low Temperature for Electrochemical Sensing Trinitrotoluene. *Anal Chem.* 2015, *87*(23), 11803-11811.

Chen, J.; Wu, W.; Zhang, F. et al. Graphene quantum dots in photodynamic therapy. *Nanoscale Adv.* 2020, *2*, 4961-4967.

Cheng, D.; Yang, L.; Li, X. et al. An Electrochemical DNA Sensing Platform Using Carboxyl Functionalized Graphene as the Electrode Modified Material. *J. Electrochem. Soc.* 2017, *164*(6), H345-H351.

Costantini, G.; Rastelli, A.; Manzano, C. et al. Universal shapes of self-organized semiconductor quantum dots: Striking similarities between InAs/GaAs(001) and Ge/Si(001). *Appl. Phys. Lett.* 2004, *85*(23), 5673–5675.

Crintea, A.; Dutu, A. G.; Sovrea, A. et al. Nanocarriers for Drug Delivery: An Overview with Emphasis on Vitamin D and K Transportation. *Nanomaterials* 2022, *12,* 1376, 1-26.

Cui, F.; Ji, J.; Sun, J. et al. A novel magnetic fluorescent biosensor based on graphene quantum dots for rapid, efficient, and sensitive separation and detection of circulating tumor cells. *Anal. Bioanal. Chem.* 2019, *411*(5), 985–995.

Demirdogen, B.C. Theranostic potential of graphene quantum dots for multiple sclerosis. *Mult. Scler. Relat. Disor.* 2022, *68*, 104232, 1-3.

Ding, H.; Zhang, F.; Zhao, C. et al. Beyond a Carrier: Graphene Quantum Dots as a Probe for Programmatically Monitoring Anti-Cancer Drug Delivery, Release, and Response. *ACS App. Mater. Interfaces* 2017, *9*(33), 27396–27401.

Dong, J.; Wang, K.; Su., L, et al. Application of graphene quantum dots for simultaneous fluorescence imaging and tumor-targeted drug delivery. *Sensors and Actuators B Chem.* 2018, *256*, 616-623.

Drbohlavova, J.; Adam, V.; Kizek, R. et al. Quantum dots - characterization, preparation and usage in biological systems. *Int. J. Mol. Sci.* 2009, *10*(2), 656–673.

Joshi, P. N.; Kundu, S.; Sanghi, S. K. et al. Graphene Quantum Dots - From Emergence to Nanotheranostic Applications. Smart Drug Delivery Systems. Edt. by A. D. Sezer, InTechOpen Publisher, 2016.

Faridbod, F.; Sanati, A. L. Graphene Quantum Dots in Electrochemical Sensors/Biosensors. *Curr. Anal. Chem.* 2019, *15*(2), 103-123.

Forte, E.; Fiorenza, D.; Torino, E. et al. Radiolabeled PET/MRI Nanoparticles for Tumor Imaging. *J. Clin Med.* 2019, *9*(1), 89,1-21.

Ganganboina, A. B.; Doong, R. A. Graphene Quantum Dots Decorated Gold-Polyaniline Nanowire for Impedimetric Detection of Carcinoembryonic Antigen. *Sci. Rep.* 2019, *9*, 7214,1-11.

Ghafary, S. M.; Nikkhah, M.; Hatamie, S. et al. Simultaneous Gene Delivery and Tracking through Preparation of Photo-Luminescent Nanoparticles Based on Graphene Quantum Dots and Chimeric Peptides. *Sci. Rep.* 2017, *7*(1), 9552,1-14.

Hui, L.; Huang, J.; Chen, G. et al. Antibacterial Property of Graphene Quantum Dots (Both Source Material and Bacterial Shape Matter). *ACS App. Mater. Interfaces* 2016, *8*(1), 20-25.

Jin, Z.; Owour, P.; Lei, S. et al. Graphene, graphene quantum dots and their applications in optoelectronics. *Curr. Opin. Colloid Interface* 2015, *20*(5-6), 439-453.

Kalkal, A.; Pradhan, R.; Kadian, S. et al. Biofunctionalized Graphene Quantum Dots Based Fluorescent Biosensor toward Efficient Detection of Small Cell Lung Cancer. *ACS Appl. Bio. Mater.* 2020, *3*(8), 4922–4932.

Ke, C.-C.; Yang, Y.-C; Tseng, W.-L. Synthesis of Blue-, Green-, Yellow-, and Red-Emitting Graphene-Quantum-Dot-Based Nanomaterials with Excitation-Independent Emission. *Part. Part. Syst. Charact.* 2016, *33*: 132-139.

Khodadadei, F.; Safarian, S.; Ghanbari, N. Methotrexate-loaded nitrogen-doped graphene quantum dots nanocarriers as an efficient anticancer drug delivery system. *Mater. Sci. Eng. C Mater. Biol. Appl.* 2017, *79*, 280–285.

Lees, E. E.; Gunzburg, M. J.; Nguyen, T. L. et al. Experimental Determination of Quantum Dot Size Distributions, Ligand Packing Densities, and Bioconjugation Using Analytical Ultracentrifugation. *Nano Lett.* 2008, *8* 2883–2890.

Lee, J. S., Youn, Y. H., Kwon, I. K. et al. Recent advances in quantum dots for biomedical applications. *J. Pharm. Investig*. 2018, *48*, 209–214.

Lin, S.; Xie, X.; Patel, M. R. et al. Quantum dot imaging for embryonic stem cells. *BMC Biotechnol*. 2007, *7*, 67,1-10.

Liu, Y.; Zhao, C.; Sabirsh, A. et al. A Novel Graphene Quantum Dot-Based mRNA Delivery Platform. *ChemistryOpen* 2021, *10*(7), 666–671.

Lo, P. Y.; Lee, G. Y.; Zheng, J. H. et al. GFP Plasmid and Chemoreagent Conjugated with Graphene Quantum Dots as a Novel Gene Delivery Platform for Colon Cancer Inhibition In Vitro and In Vivo. *ACS App. Bio. Mater.* 2020, *3*(9), 5948–5956.

Malode, S. J.; Shetti, N. P. Chapter 10 - Carbon dots-based electrochemical sensors, Carbon Dots in Analytical Chemistry, In:Detection and Imaging; Kailasa, S.K. and Hussain, C.M. Eds., Elsevier, 2023, 109-136.

Mansuriya, B. D.; Altintas, Z. Graphene Quantum Dot-Based Electrochemical Immunosensors for Biomedical Applications. *Mater.* 2019, *13*(1), 96,1-30.

Markovic, Z. M.; Ristic, B. Z.; Arsikin, K. M. et al. Graphene quantum dots as autophagy-inducing photodynamic agents. *Biomater.* 2012, *33*(29), 7084–7092.

Moghadam, N. B.; Avatefi, M.; Karimi, M. et al. Graphene family in cancer therapy: recent progress in cancer gene/drug delivery applications. *J. Mater. Chem. B.* 2023, *11*, 2568-2613.

Mollarasouli, F.; Serafín, V.; Campuzano, S. et al. Ultrasensitive determination of receptor tyrosine kinase with a label-free electrochemical immunosensor using graphene quantum dots-modified screen-printed electrodes. *Anal. Chim. Acta.* 2018, *1011*, 28–34.

Mueller, M. L.; Yan, X.; McGuire, J. A. et al. Triplet States and Electronic Relaxation in Photoexcited Graphene Quantum Dots. *Nano Lett.* 2010, *10*(7), 2679-2682.

Naresh, V.; Lee, N. A. Review on Biosensors and Recent Development of Nanostructured Materials-Enabled Biosensors. *Sensors*. 2021, 21, 1109.

Nikazar, S.; Sivasankarapillai, V. S.; Rahdar, A. et al. Revisiting the cytotoxicity of quantum dots: an in-depth overview. *Bioph. Rev.* 2020, *12*(3), 703–718.

Niu, X.; Zhong, Y.; Chen, R. et al. A "turn-on" fluorescence sensor for Pb2+ detection based on graphene quantum dots and gold nanoparticles. *Sens. Actuators B: Chem.* 2018, *255*(2), 1577-1581.

Nigam, P.; Waghmode, S.; Lois, M. et al. Graphene quantum dots conjugated albumin nanoparticles for targeted drug delivery and imaging of pancreatic cancer. *J. Mater. Chem B.* 2014, *2*, 3190-3195.

Nsairat, H.; Khater, D.; Sayed, U. et al. Liposomes: structure, composition, types, and clinical applications. *Heliyon* 2022, *8*(5), e09394,1-15.

Perini, G.; Palmieri, V.; Ciasca, G. et al. Unravelling the Potential of Graphene Quantum Dots in Biomedicine and Neuroscience. *Int. J. Mol. Sci.* 2020, *21*, 3712,1-25.

Prabhakar, A. K.; Ajith, M.; Ananthanarayanan, A. et al. Ball-milled graphene quantum dots for enhanced anti-cancer drug delivery. *OpenNano* 2022, *8*, 100072,1-11.

Qi, L.; Pan, T.; Ou, L. et al. Biocompatible nucleus-targeted graphene quantum dots for selective killing of cancer cells via DNA damage. *Commun. Biol.* 2021, *4*(214), 1-12.

Raj, M. A.; Jhon, S. A. Chapter 1 - Graphene-Modified Electrochemical Sensors, Graphene-Based Electrochemical Sensors for Biomolecules, Micro and Nano Technologies, Pandikumar, A. and Rameshkumar, P., Eds, Elsevier Publisher, 2019, 1-41.

Ratemi, E.; Sultana Shaik, A.; Al Faraj, A. et al. Alternative approaches for the treatment of airway diseases: focus on nanoparticle medicine. *Clin. Exp. Allergy* 2016, *46*, 1033–1042.

Roushani, M.; Valipour, A.; Bahrami, M. The potentiality of the functionalized nitrogen and thiol-doped graphene quantum dots (GQDs-N-S) to stabilize the antibodies in the designing of human chorionic gonadotropin immunosensor. *Nanochem. Res.* 2019, *4*(1), 20-26.

Saffioti, N. A.; Cavalcanti-Adam, E. A.; Pallarola, D. Biosensors for Studies on Adhesion-Mediated Cellular Responses to Their Microenvironment. *Front. Bioeng. Biotechnol.* 2020, *8*, 597950,1-21.

Samimi, E.; Karami, P.; Ahar, M. J. A Review on Aptamer-Conjugated Quantum Dot Nanosystems for Cancer Imaging and Theranostic. *J. Nanomed. Res.* 2017, *5*(3), 00117,1-8.

Savas, S.; Altintas, Z. Graphene Quantum Dots as Nanozymes for Electrochemical Sensing of *Yersinia enterocolitica* in Milk and Human Serum. *Materials* 2019, *12,* 2189,1-14.

Senel, B.; Demir, N.; Büyükköroğlu, G. et al. Graphene quantum dots: Synthesis, characterization, cell viability, genotoxicity for biomedical applications. *Saudi Pharm. J.* 2019, *27*(6), 846–858.

Shaari, N.; Kamarudin, S. K.; Bahru, R. Carbon and graphene quantum dots in fuel cell application: An overview. *Int. J. Energy Res.* 2021, *45*, 1396–1424.

Sui, X.; Luo, C.; Wang, C. et al. Graphene quantum dots enhance anticancer activity of cisplatin via increasing its cellular and nuclear uptake. *Nanomedicine* 2016, *12*(7), 1997–2006.

Sun, H.; Wu, L.; Wei, W. et al. Recent advances in graphene quantum dots for sensing. *Mater. Today* 2013, *16*(11):433–442.

Szymanski, C. J.; Yi, H.; Liu, J. L. et al. Imaging intracellular quantum dots: fluorescence microscopy and transmission electron microscopy. *Methods Mol. Biol.* 2013, *1026*, 21–33.

Tabish, T. A.; Scotton, C. J.; Ferguson, D. C. J. et al. Biocompatibility and toxicity of graphene quantum dots for potential application in photodynamic therapy. *Nanomedicine* 2018, *13*(15), 1923–1937.

Tan, L.; Shen, X.; He, Z. et al. The Role of Photodynamic Therapy in Triggering Cell Death and Facilitating Antitumor Immunology. *Front. Oncol.* 2022, *12*, 863107,1-9.

Texier, I.; Josser, V. In vivo imaging of quantum dots. *Methods Mol. Biol.* 2009, *544*, 393–406.

Tian, Z.; Yao, X.; Ma, K. et al. Metal-Organic Framework/Graphene Quantum Dot Nanoparticles Used for Synergistic Chemo- and Photothermal Therapy. *ACS Omega* 2017, *2*(3), 1249–1258.

Xiao, S.; Zhou, D.; Luan, P. et al. Graphene quantum dots conjugated neuroprotective peptide improve learning and memory capability. *Biomater.* 2016, *106*, 98–110.

Xu, W.; Lee, T. W. Recent progress in fabrication techniques of graphene nanoribbons. *Mater. Horiz.* 2016, *3*, 186-207.

Wagner, A. M.; Knipe, J. M.; Orive, G. et al. Quantum dots in biomedical applications. *Acta Biomaterialia* 2019, *94*, 44–63.

Wang, X.; Sun, X.; Lao, J. et al. Multifunctional graphene quantum dots for simultaneous targeted cellular imaging and drug delivery. *Colloids Surf. B, Biointerfaces* 2014, *122*, 638–644.

Wang, J.; Liu, G.; Leung, K. C. et al. Opportunities and Challenges of Fluorescent Carbon Dots in Translational Optical Imaging. *Curr. Pharm. Des.* 2015, *21*(37), 5401–5416.

Wang, H.; Mu, Q.; Wang, K. et al. Nitrogen and Boron Dual-Doped Graphene Quantum Dots for Near-Infrared Second Window Imaging and Photothermal Therapy. *App. Mater. Today* 2019, *14*, 108–117.

Wei, Z.; Yin, X.; Cai, Y. et al. Antitumor effect of a Pt-loaded nanocomposite based on graphene quantum dots combats hypoxia-induced chemoresistance of oral squamous cell carcinoma. *Int. J. Nanomedicine* 2018, *13*, 1505–1524.

Yan, Q.; Yang, Y.; Tan, Z. et al. A label-free electrochemical immunosensor based on the novel signal amplification system of AuPdCu ternary nanoparticles functionalized polymer nanospheres. *Biosens. Bioelectron.* 2018, *103*, 151–157.

Yang, Y.; Liu, Q.; Liu, Y. et al. A novel label-free electrochemical immunosensor based on functionalized nitrogen-doped graphene quantum dots for carcinoembryonic antigen detection. *Biosens. Bioelectron.* 2017, *90*, 30-38.

Ying, P.; Liang, T.; Du, Y. et al. Thermal transport in planar -hybridized carbon allotropes: A comparative study of biphenylene network, pentaheptite and graphene. *Int. J. Heat Mass. Trans.* 2022, *183*, 122060,1-12.

Yu, C.; Long, Z.; Qiu, Q et al. Graphene quantum dots-based targeted nanoprobes detecting drug delivery, imaging, and enhanced chemotherapy of nasopharyngeal carcinoma. *Bioeng. Transl. Med.* 2021, *7*(2), e10270,1-14.

Yusuf, A.; Almotairy, A. R. Z.; Henidi, H. et al. Nanoparticles as Drug Delivery Systems: A Review of the Implication of Nanoparticles' Physicochemical Properties on Responses in Biological Systems. *Polymers* 2023, *15*(7), 1596,1-26.

Zhao, C., Song, X., Liu, Y. et al. Synthesis of graphene quantum dots and their applications in drug delivery. *J. Nanobiotechnol*. 2020, *18*, 142,1-32.

Zhou, Z.; Song, J.; Nie, L. et al. Reactive oxygen species generating systems meeting challenges of photodynamic cancer therapy. *Chem. Soc. Rev.* 2016, *45*, 6597-6626.

Zia-ur-Rehman, M.; Qayyum, M. F.; Akmal, F. et al. Chapter 7 - Recent Progress of Nanotoxicology in Plants. Editor(s): D. K. Tripathi, P. Ahmad, S. Sharma, D. K. Chauhan, N. K. Dubey, Nanomaterials in Plants, Algae, and Microorganisms, Academic Press, 2018, 143-174.

Zumpano, R.; Polli, F.; D'Agostino, C. et al. Nanostructure-Based Electrochemical Immunosensors as Diagnostic Tools. *Electrochem.* 2021, *2*, 10-28.

Chapter 7

Graphene Quantum Dots: Investigations and Applications

Priya Rani[1]
and Sunita Srivastava[2]
[1]Department of Physics, Guru Jambheshwar University of Science and Technology, Hisar, India
[2]Department of Physics & Astrophysics, Central University of Haryana, Mahendergarh, India

Abstract

Graphene Quantum Dots (GQDs) are of particular interest in semiconductor applications and bio-nanotechnology. These are minuscule fragments of graphene sheets having carbon atoms arranged in a hexagonal framework with a size comprising 10 to 100 atoms and hence exhibit astounding electrical and optical properties compared to graphene sheets as a consequence of quantum confinement. For their use in real-world applications, it is extremely important to analyze their regulating parameters and consequently optimize their band gaps. Changing the size is one of the well-known approaches in engineering their bandgaps, and one may tailor the spectacular characteristics of these materials by incorporating heteroatoms into them and by surface modifications with certain groups in order to meet the customized requirements of the envisioned application. In this chapter, the particular requirements of some applications and the methods involved in modifying the properties of GQDs for specific applications have been discussed. A brief description of some of the problems that must be addressed in order to reap the beneficial effects of GQDs on a wide scale has also been discussed.

In: Graphene Quantum Dots (GQDs)
Editor: Aleksey Kuznetsov
ISBN: 979-8-89113-638-0

1. Introduction

The most recent acquisition of the group of nanocarbon compounds is the graphene quantum dot (GQD). It is a 0D graphene material with an average lateral size of less than 10 nm with an atom thickness (usually 1 or 2 layers, around 2 nm thick). GQDs possess unique properties because they incorporate a graphene honeycomb throughout the dots. When compared to its 2D counterpart, graphene sheet, GQD has some distinct advantages that open entirely novel possibilities for applications. These advantages include bandgap opening resulting from quantum confinement, excellent dispersibility, a greater abundance of active sites, superior tunability in physico-chemical properties, and size similar to that of biological molecules [1]. Usually, everyone gets confused between GQD and another class of 0D carbon compounds (carbon dots, CDs). These are different in morphology and have very different properties. CDs possess amorphous structures having sp^3 hybridization along with crystalline structures in some cases with sp^2 hybridization, while GQDs are pure crystalline structures with sp^2 hybridization all over the structure [2]. In contrast to CDs, whose bandgap structures are determined by surface energy traps, GQDs' electronic bandgap structures are significantly influenced by the quantum confinement effect [3]. Because of these variations, some techniques or applications could possibly be effective for one but not the other.

Among other QDs, group IV-VI QDs, such as CdSe, PbTe, PbSe, CdS, HgTe, and ZnSe, have gained a lot of popularity in recent years due to their wide range of applications, which include optoelectronics, solar cells, light-emitting diodes, and bio-imaging [4, 5, 6, 7]. The capability of adapting the optical and electrical properties of quantum dots allows these numerous applications [8]. The primary drawbacks of these colloidal QDs are their inherent toxicity and poor stability, which prohibit their usage in many applications. On the other hand, GQDs are less toxic and exhibit size dependent electronic and optical properties along with the exceptional properties of graphene, hence preferable for such applications. A brief history and development of GQDs from graphite is discussed in the next section.

1.1. From Graphite to Graphene Quantum Dots

A breakthrough in the scientific community is currently taking place as researchers reveal significant techniques to pattern and describe materials at the nanoscale length. This technology may be regarded as a key mechanization of the twenty-first century and is referred to as Nanotechnology. Among all materials, carbon has undergone substantial exploration since the seventeenth century and has endured an unremitting increase ever since [9, 10, 11]. The research proliferation started after Brodie synthesized graphite oxide, sometimes known as "graphon," for assessing the molecular weight of graphite [12]. In 1918, Kohlschütter and Haenni proceeded to demonstrate the peculiarities of graphite oxide [13]. As a consequence, Bernal [14] evaluated its psychological behaviour using single-crystal diffraction in 1924. As characterization techniques were further improved, Wallace subsequently studied the electronic characteristics of graphite and developed the notion of graphene, which he eventually referred to as "a wonder material" [15]. Later, transmission electron microscopy (TEM) was effectively utilised to explore several graphite surfaces [16]. The process of manufacturing graphene oxide, which was first invented in 1918, was extremely risky. Therefore, Hummers and Offeman later developed another safe, trustworthy, cost-effective method [17]. A few years later, Boehm documented the development of a single-layer graphite sheet [18]. Then unimaginable work was carried out by theoretical physicists DiVincenzo and Mele by developing a massless Dirac equation for the graphite host in 1984 [19]. The scientific community had an epiphany moment in 1985 when Smally discovered buckminsterfullerene (C60). A year later, Smally received the Nobel Prize for his astounding investigation [20]. Mouras et al. used the word 'graphene' to describe single sheets of graphite in 1987, although recognised that mathematical computations had remained the subject of imagination at the time [21]. After more than 10 years, Geim and Novoselov's outstanding experiment captured a pivotal point in the development of carbon materials [22]. They synthesised monolayer graphene, which inspired the whole scientific community to look into two-dimensional materials in order to develop new and more sophisticated technological applications. Their discoveries on carbon materials once more perceived their path in 2010 when Geim and Novoselov were awarded the Nobel Prize for their meritorious experimental work.

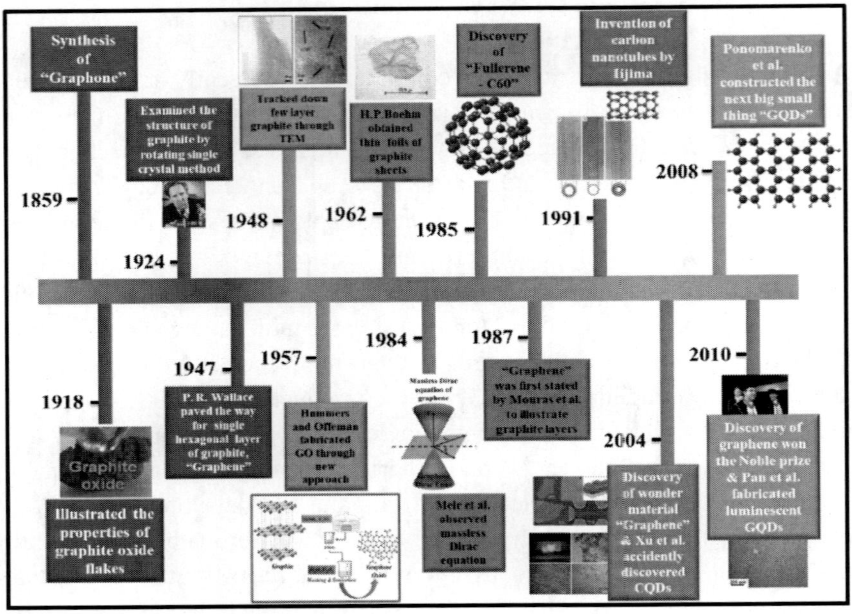

Figure 1. Development journey of graphene quantum dots from graphite [30].

Since the outbreak's identification, graphene has been a focus of research because of its excellent electrical [23], mechanical [24], optical [25], and other properties that may be utilised for a wide range of applications, including organic light emitting diode (OLED), lithium-ion batteries, ultra-capacitors, water desalination, and others [26]. Graphene behaves as a semi-metal due to an extremely small overlap between its conduction band and valence band [27]. Aside from these intriguing characteristics, graphene noteworthy drawbacks include accumulation in the solvents, low absorptivity, poor dispersion in solvents, and zero band gaps [27]. Due to graphene's zero bandgap, a property exhibited by half-metallic substances, its application in the field of electronic devices and semiconductors is constrained [28]. Different strategies, including structural modification and the addition of adjacent atoms, were developed to get around graphene's restrictions. One promising method to engineer bandgaps in graphene is to modify graphene two-dimensional (2D) sheet into quantum dots (QDs) with zero-dimension (0D). Once the size of graphene sheets is shrunk to the nanoscale by terminating their free bonds with hydrogen, graphene quantum dots (GQDs) are formed. In 2008, Ponomarenko and Geim developed "GQDs" [29]. These insignificant dots are quickly emerging as the future of research society. The development history of GQDs from graphite is represented in Figure 1. GQDs possess band gaps in contrast to graphene sheets allowing them to have special electrical and optoelectronic characteristics. Additionally, GQDs have a feature of size-dependent optical properties, making them suitable for a wide range of applications.

2. Physiochemical Properties of Graphene Quantum Dots

Nanoscale graphene sheets are called GQDs. Their morphology, Bohr radius and the quantum confinement effect have been discussed in this section in order to explain the optical characteristics of GQDs. These crucial terms are thought to be associated with the changes in the photoluminescence, electrical, and physical features of GQDs. Let's take a quick look at these words to improve the grasp of the fundamental characteristics of GQDs.

2.1. Morphology and Structure

GQDs consist of a graphene core with sp^2 hybridized carbon atoms with their dangling bonds terminated by hydrogen atoms. However, experimentally fabricated GQDs always have numerous oxygen reactive groups present both at their edge locations and on the basal planes. Epoxy groups are generally found in the basal plane of GQDs [31, 32] while other groups, such as hydroxyl, carboxyl, and carbonyl, are attached to the edge sites of GQDs [33]. GQDs usually have circular or elliptical forms; however, experimental research showed that they can also be polygonal [34, 35]. By adjusting synthesis parameters like reaction period, temperature, contacting medium, etc., one can regulate the size and shapes of GQDs [36]. The graphene sheet is primarily sliced into two distinct crystallographic orientations for the fabrication of GQDs, resulting in two different kinds of edge structures for GQDs: armchair and zigzag [37]. These two structures are represented in Figure 2.

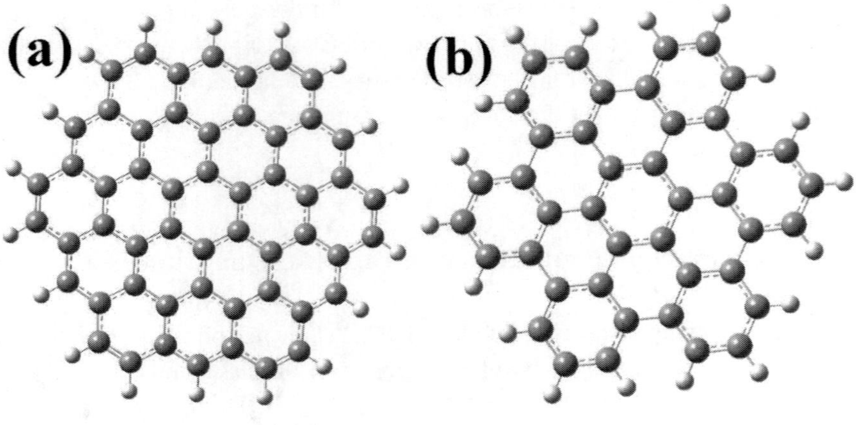

Figure 2. (a) Zigzag and (b) armchair structures of GQDs.

According to Kim et al., polygonal-shaped GQDs exhibit mostly the armchair edges, whereas circular/elliptical-shaped GQDs exhibit both zigzag and armchair edges [38]. GQDs normally have sixfold symmetry and exist as crystal; however, any kind of doping or defect might lead to a variation in the lattice spacing, which could destroy the periodicity [36]. When describing the kind of crystal structure, edge state, and imperfections in GQDs, the D and G bands in the Raman spectra are frequently employed [39]. The D band at 1350 cm^{-1} is largely produced by the sites at armchair edge [40]. The G band, which is believed to be a distinguishing peak in the Raman spectrum of a nanocarbon material, is divided into the G^- and G^+ bands in accordance with the structure of the carbon nanotube [40]. Additionally, it has been shown that the G^+ and G^- bands are similarly separated in the Raman spectra of zigzag GNRs. The G^+ and G^- bands, respectively, originate in the basal plane of zigzag GNRs. Additionally, the functional groupings and edge geometry of GQDs have a considerable impact on their optical properties [31, 32, 41].

GQDs having holes inside their structures are also trending now a day which is called holey GQDs. Mumtaz Ali et al. studied some of holey GQDs and found that introducing holes inside the GQDs can possibly enhance the various properties of GQDs [42]. They introduced pores in the GQDs to improve the electrocatalytic activity. In the dye-sensitized solar cell, the counter electrode with a porous composite structure demonstrated 78% improved photovoltaic performance (in comparison to unmodified graphene), which is comparable to Pt electrode performance [42]. Emerging electrocatalytic applications including supercapacitors, water splitting, and battery electrodes can make use of the suggested 3D porous orientation [43].

Structures of holey GQDs with their pores passivated with hydrogen and oxygen atoms are shown in Figure 3.

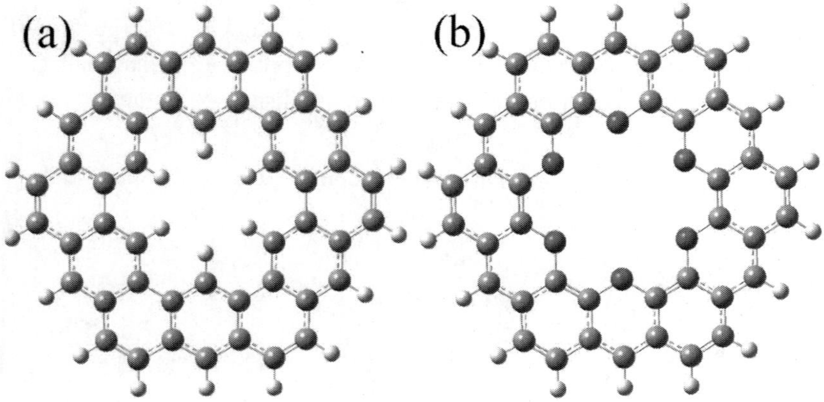

Figure 3. Holey GQDs having inner part passivated (a) with H-atoms, (b) with O-atoms.

2.2. Bohr Radius

Although quantum dots have the same structural characteristics as the parent molecules, they also have special electrical and optical properties depending on their size. When these nanostructures become smaller than a specific radius, the quantum size effect takes place [44]. This radius is known as Bohr radius of QDs. The Bohr radius is defined as the distance between the electron excited to the conduction band caused by the absorption of a photon by the semiconductor and hole left behind in the valence band [45]. Graphene has infinite Bohr radius [46].

2.3. Quantum Confinement Effect

When the quantum dot size decreases and gets closer to the bulk exciton's Bohr radius, the quantum confinement effect comes into play [44]. Depending on the confinement size, three distinct types of structures may be distinguished: a quantum well (QW), a quantum wire (QWR), and a quantum dot (QD) [47]. While the exciton (combination of an electron and a hole) is allowed to move in all three directions in a QW, the material dimension is diminished in two directions in a QWR, limiting the exciton to only one of the three directions. In a QD, each of three dimension is constrained, obstructing the free moment of excitons in all the three directions [48]. The confinement of the exciton in these three configurations is shown in Figure 4. This confinement effect modifies the exciton's nature, providing quantum dots with distinct optical and electrical features which are discussed in the next section.

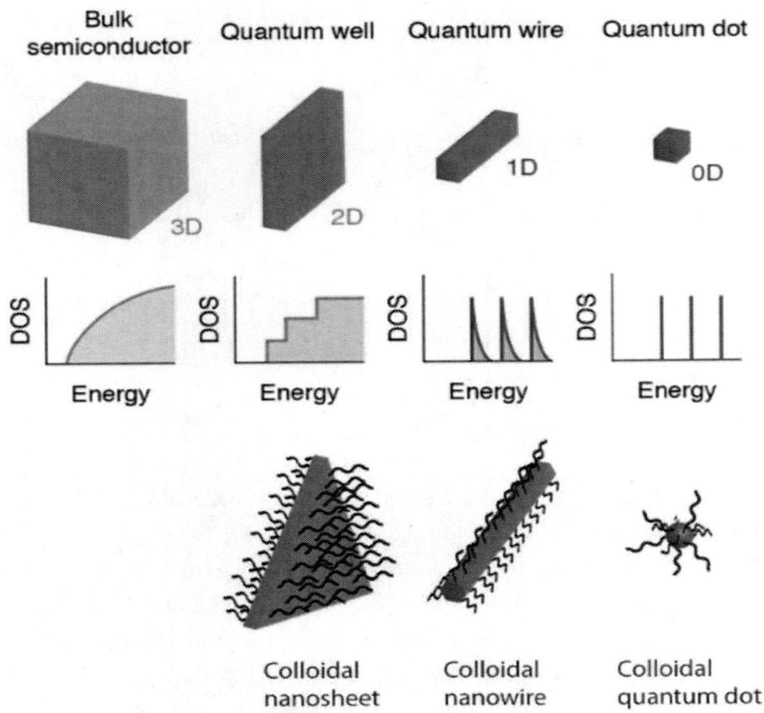

Figure 4. Systematic representation of change in energy levels with the reduced dimensionality (visual representation of quantum confinement) [49].

2.4. Optical Properties of GQDs

Due to their outstanding optical properties, GQDs have been broadly used for cutting-edge research and sophisticated applications since their creation [31, 50, 51, 52]. The majority of analyses of the optical characteristics of GQDs are based on photoluminescence (PL) spectroscopy and UV-vis absorption measurements [53]. Two distinct kinds of electronic transitions, specifically the π-π* and n-π* transitions, are involved to unravel the UV-vis absorption spectra of GQDs. The absorption curve in the range 230-270 nm is mainly due to π-π^* transition, and the broader ends in the range from 300-600 nm are due to n-π^* transitions which are caused by surface states, defects, functional groups, doping, etc. in GQD [54]. The optical bandgap of GQDs determines their UV-vis absorption, which itself is entirely controlled by the GQD fundamental characteristics, such as size, shape, etc., as well as the additional parameters, such as surface modification, doping, and defects [55]. With the diminution in the size of GQDs, the blue shift of the absorption peak becomes evident, indicating the quantum confinement effect (QCE) effect [38, 56]. After oxygen functionalization, the broadening of the absorption spectra associated to the n-π^* transition has also been noted [31]. PL emission is another intriguing characteristic attribute of GQDs, similar to UV-vis absorption. PL is essentially the emission of photons caused by excitons relaxing to lower energy levels. Depending on the inherent features and surface states of the GQDs as they are activated with UV-vis light, PL emission of different colours ranging from blue to red is recognised [32, 55, 52, 57, 58]. Additionally, the synthesis technique, excitation wavelength, solvent, pH, and other factors have an impact on the colours of PL emission in GQDs [32, 51]. The QCE, or the size of the GQDs, primarily explains the source of the PL emission in GQDs [52, 59] The QCE mechanism states that as GQD size is decreased, the energy bandgap widens, which causes the PL peak of GQDs to shift to a shorter wavelength, [51] as depicted in Figure 5.

Additionally, another reason of the excitation-dependent PL emission in GQDs is owing to the close proximity of energy levels, referred to as inherent states, appearing inside the energy bandgap owing to the uneven distribution of particle sizes of GQDs [52, 60]. Excitation-dependent PL emission in GQDs was demonstrated by Zhu et al. [61]. They also proved that as the wavelength of excitation changes from 400 nm to 540 nm, the peak of the emission moves from 515 to 570 nm, with the peak at 515 nm being the strongest [61]. Furthermore, the research group asserted that excitation-independent PL is feasible with GQDs due to their consistent size and emissive locations [51]. PL emission in GQDs, which frequently displays green emission, is attributed to the abundance of oxygen functional groups in these structures. Although the QCE and surface states assist in clarifying the process of the PL emission in GQDs, certain aspects of the PL performance in GQDs remain unclear [62]. Along with PL emission in GQDs, a crucial criterion for identifying GQDs as a fluorescent material, is PL quantum yield (QY), which is defined as the ratio of emitted photons by the material to the photons absorbed at a specific excitation wavelength [63]. The QY values of GQDs have been reported to be in the range of 2.9 to 83%, [61, 64, 65], which is extremely high and establishes GQDs as suitable luminescent materials, especially for physiological monitoring and bio-imaging purposes [66]. Generally, in order to use GQDs in various applications, their optical properties need to be modified using specific techniques. Several strategies are currently being used to modify the properties of GQDs, some of them are explained below.

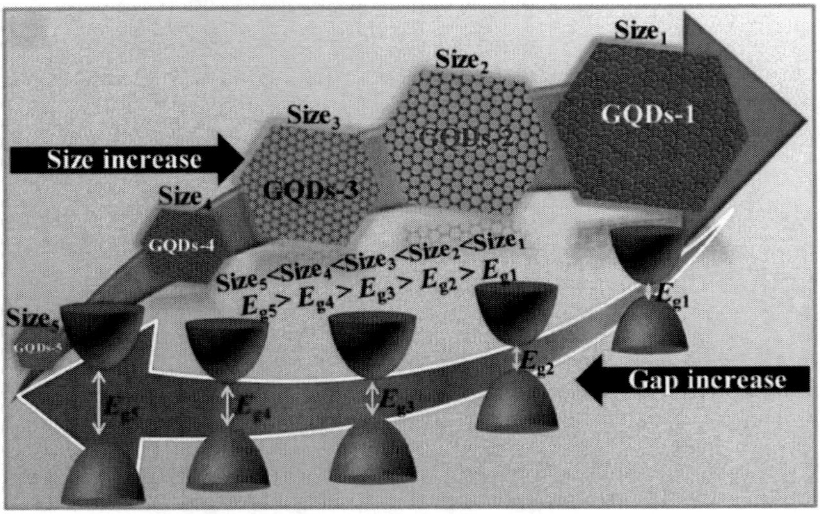

Figure 5. Size dependent energy band gap and photoluminescence of GQDs [34].

3. Modification of Properties of Graphene Quantum Dots

As GQDs possess the properties of both QDs and graphene, they have very distinctive optical characteristics. But to use them in practical applications, their properties are required to be modified accordingly. The only genuine uses of pristine GQD are for electrical gadgets, additionally, it is difficult to synthesize and distribute GQDs [34]. Actually, chemical groups, functionalized groups, imperfections, dopants, and their cohabitation make GQDs intriguing as well as advantageous for practical applications. Here, a number of strategies are considered that make GQDs appealing materials.

3.1. Changing Size and Shape

The core characteristic of QDs is the dependence of the electrical and optical properties of GQDs on their shape and size. This effect is referred to as the quantum confinement effect and has also been shown by GQDs [67, 68, 69].

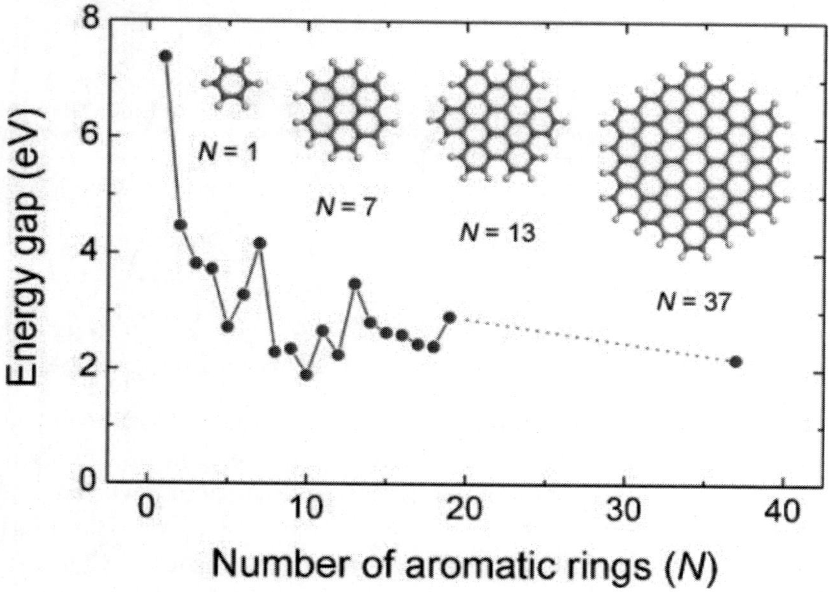

Figure 6. Change in band gap of GQDs with their size [70].

The bandgap between the valence and conduction bands continues to narrow down as GQD size increases as represented in Figure 6. The bandgap of GQDs decreases from 7.2 eV [67] to 2.85 eV [70] when the number of ring of hybridized C-atoms in them increases from 1 to 37. Their shape and edge structures also influence how GQDs behave. GQDs of the same size with various edge configurations, such as zigzags and armchairs, are discovered to have various bandgaps [71]. This indicates that by adjusting the GQDs size and shape, their unique characteristics may be tailored to a particular application.

3.2. Passivation

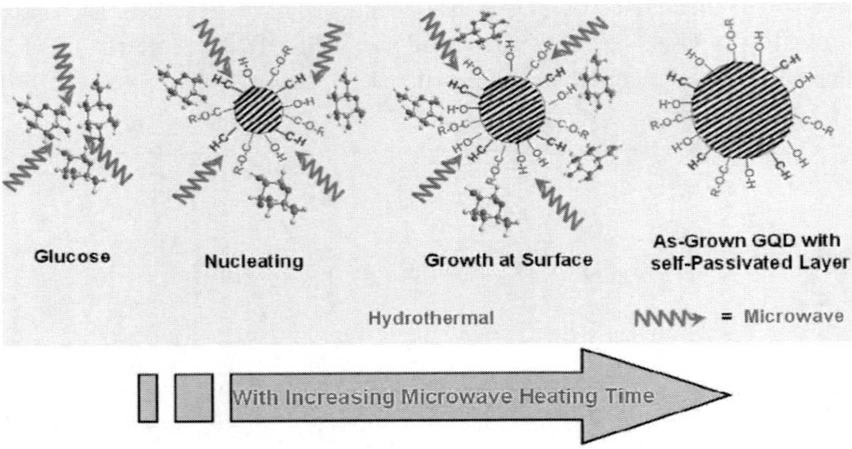

Figure 7. The self-passivated GQDs developed by hydrothermal approach [57].

Passivation is a strategy which is used to terminate the dangling bond at the edge of GQDs. Passivation also serves as an important strategy used to modify various properties of GQDs [68]. Thus, Tang et al. prepared GQDs using hydrothermal approach and reported that the self-passivated layer on the GQDs' surface resulted in distinctive optical characteristics of GQDs [57]. These GQDs possessed the shortest deep ultraviolet (DUV) emission wavelength amongst the various solution-based QDs at 4.1 eV. This is attributed to self-passivated layers on the surfaces of GQDs. The as-grown GQDs are represented in Figure 7. Cai et al. demonstrated that self-passivated single-layered GQDs can be used to detect Cr(VI) ions. The passivating layer on the surface of GQDs helps to interact with Cr(VI) and make a complex which quenches the fluorescence and red-shifts the UV adsorption which indicates the presence of Cr(VI) ions in the solution [72]. The passivation of single-layered GQDs with electronegative elements such as F also results in the red shift in the UV absorption spectra and emission spectra owing to decrement in the band gap after this passivation [68], which results because of the introduction of energy levels related to F-atom inside the band gap of GQDs which cause a reduction in the band gap and corresponding change in the optical properties. This effect is represented in Figure 8.

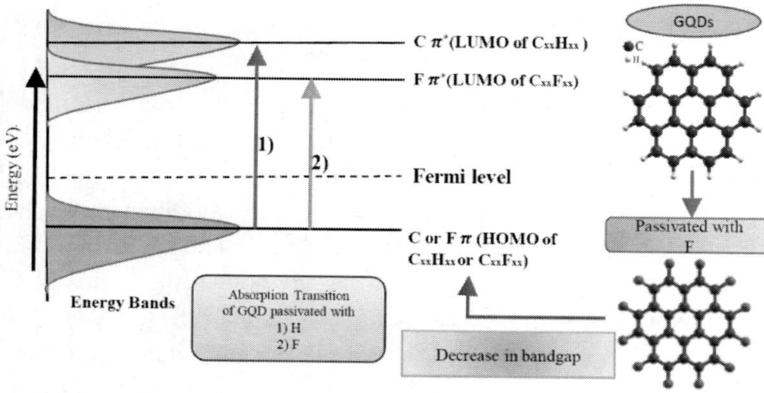

Figure 8. Introduction of F-atom related energy levels in between the band gap of GQD on passivation them with F-atoms [68].

Table 1. Absorption energy (E_{abs}) and wavelength (λ) of dominant excitation state of GQDs [68]

GQDs	Dominant Excitation	E_{abs} (eV)	λ (nm)	O. S. (f)
$C_{24}H_{12}$	S3	4.04	300.88	0.96
	S4	4.13	300.25	0.94
$C_{24}F_{12}$	S3	3.94	314.66	0.95
	S4	3.95	314.64	0.94
$C_{42}H_{18}$	S5	3.46	358.12	1.06
	S6	3.46	358.09	1.06
$C_{42}F_{18}$	S3	3.33	372.88	0.94
	S4	3.33	372.17	0.93
$C_{54}H_{18}$	S3	2.87	432.28	1.39
	S4	2.87	432.27	1.39
$C_{54}F_{18}$	S3	2.74	452.05	1.35
	S4	2.74	452.04	1.35

The results of the passivation by F-atoms on GQDs have also been depicted in the Table 1. This table shows the variation in the dominant absorption energy in the absorption spectra on terminating the GQDs with electronegative atoms. The oscillator strength (O. S.) of absorption peak remains the same even after passivation of GQDs because this absorption is due to the sp^2 hybridized region of C atoms within the GQDs which remains the same after passivation, because passivation only replaces the H-atoms with F-atoms and has no effect on the inner hybridization of C-atoms. It is clear from the table that the red shift observed for the absorption peaks is significant and thus passivation is a great strategy to modify the properties of the GQDs.

3.3. Heteroatom Doping

Since the discovery of GQDs, researchers have been working to modify the bandgap of GQDs using a variety of approaches to make them usable in sophisticated applications in the UV to NIR region. One such method is the doping of GQDs by heteroatoms, which greatly modifies their physical, optical, and electrical properties [73]. Here, it is explained in detail how doping with various heteroatoms modifies the optical properties of GQDs. Heteroatoms such as nitrogen (N) [74, 75, 76], sulphur (S) [77, 78], boron (B) [79, 80], and phosphorous (P) [81], as well as some others, are frequently employed for doping in GQDs. Additionally, elements such as potassium (K) [82], selenium (Se) [83], fluorine (F) [84], chlorine (Cl) [85], and others are employed as dopants in GQDs. Additionally, such atoms are also suitable for co-doping of GQDs to enhance their application potential [79, 81].

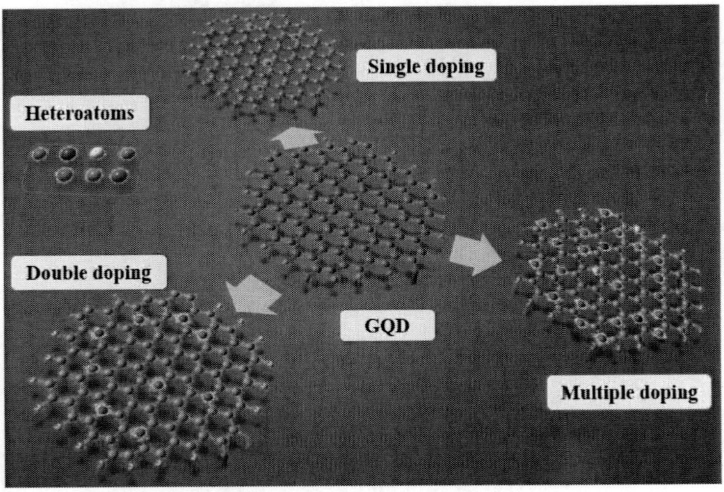

Figure 9. Various types of heteroatom doping in GQDs.

According to the nature of doping atoms, types of doping are usually classified as single, double, and multiple doping, as shown in the Figure 9. Keep in mind that the choice of the doping type has a greater impact on the characteristics of GQDs than the dopant itself. Also, synthetic approach for obtaining doped GQDs is important. The researchers used different synthesis methods for doping hetero atoms in GQDs and produced various doped GQDs with different characteristic optical properties. For instance, Tang et al. fabricated N-GQDs at ambient temperature and atmospheric pressure by combining a mixture of glucose with ammonia in a liquid media, where the ammonia serves as both a catalyst and a source of N [86]. N is incorporated into the synthesized N-GQDs as graphitic, pyrrolic, and pyridinic N, with pyridinic N making up the majority of the N component. The authors suggested that the N doping causes the band structure of GQDs to produce new energy levels. They reported certain modifications to the UV-vis absorption and PL emission spectra as a result of the band structure modification. According to Zhu et al., increasing the amount of N doped into GQDs causes an increase in their band gap [74]. As a result, the PL spectra become customised, and the PL efficiency is improved. It has been shown that the doping with O, S, and Se results in the shift of absorption and fluorescence (FL) of GQDs along with the increase in their FL QY [87]. Figure 10 represents the effect of doping by O, S, and Se atoms in C_{54} GQD on the absorption spectra. The absorption spectrum in Figure 10 (a) shows that the longest absorption wavelength is located at 432 nm in pure GQD while Figure 10 (b) shows that the longest absorption peaks of doped GQDs appear at 1042.35 nm, 750.25 nm, and 715.40 nm for $C_{52}O_2H_{18}$, $C_{52}S_2H_{18}$, and $C_{52}Se_2H_{18}$ GQDs, respectively.

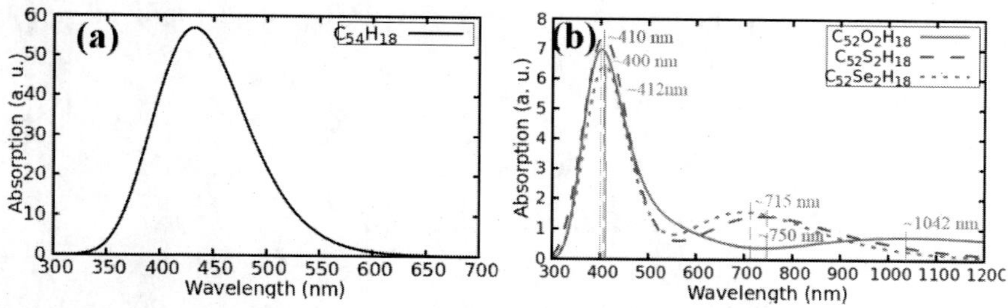

Figure 10. Absorption spectra of (a) pristine GQDs, (b) heteroatom doped C_{54} GQDs.

Also, the effect of this doping on FL spectra is depicted in Figure 11 with the corresponding QY represented in Table 2. Doping results in the shift of fluorescence peak of GQDs with significant fluorescent peak in NIR for doped GQDs. Also, the bare GQDs have negligible QY [87]. Doping of heteroatoms to GQDs significantly enhanced their QY in near infrared region (NIR) making them suitable for bioimaging applications as is clear from the Table 2.

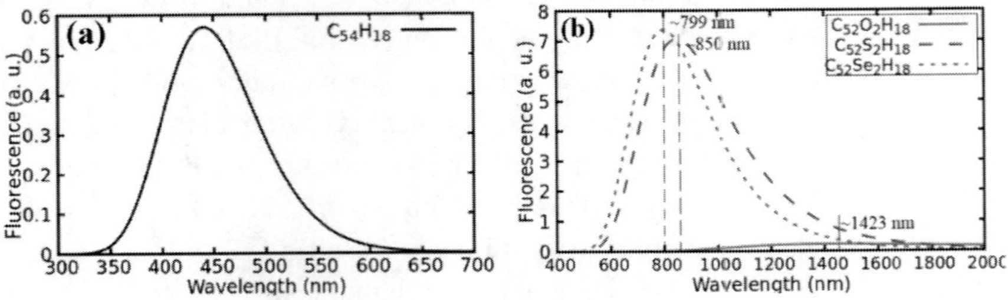

Figure 11. Emission spectra of doped C_{54} GQDs.

Table 2. Fluorescence QY of doped GQDs in NIR

GQDs	$C_{52}O_2H_{18}$	$C_{52}S_2H_{18}$	$C_{52}Se_2H_{18}$
QY (%)	1.37	25.91	24.77

The co-doping of N and Br produced the yellow emission from GQDs with 50% QY [88]. Recent report suggests that by utilising the suitable synthesis techniques, the emission QY of CQDs might be boosted up to 84% in the visible range [89]. In particular, the introduction of heteroatoms caused a corresponding shift in the UV-vis absorption of GQDs and improved various other properties notably, therefore, heteroatom doping is an excellent strategy to tailor the properties of GQDs.

3.4. Surface Modification/Functionalization

During synthesis, uncontrolled functionalization is a necessary component of GQDs which cannot be avoided, because the majority of their production procedures use chemical techniques. However, in GQDs containing dominant electron-donating/accepting groups, regulated functionalization significantly affects their extrinsic properties, including optical, chemical, and electrical properties. Synthesized GQDs always have some of oxygen and hydrogen containing groups. It should be highlighted that in GQDs, C-O and -OH groups function as the emissive PL centers [32, 51] whereas C-O-C and -COOH groups primarily act as non-radiative centers and carrier capturing centers [64, 90]. According to Jin et al., amine-functionalized GQDs have a PL emission peak that is around 30 nm redshifted when compared to naked GQDs [91]. They suggested that the bandgap in functionalized GQDs is reduced as a result of the amine groups with plenty of electrons contributing their lone pairs to increase the density of electrons in GQDs. Other group showed that when

GQDs are functionalized with aryl amines, they increase the sp^2 conjugated region of the QD and correspondingly lead to lowering of π^* levels resulting in the reduction of band gap and red shift in PL and absorption spectra [92]. GQD edge structures, in addition to GQD size and functional groups, have an impact on the PL emission of GQDs. Some researchers used the modifications of both zigzag and armchair edge states to show how the PL emission spectra of GQDs can alter [31]. Functionalization of GQDs with boronic and borinic acids also adds a remarkable change in GQDs. The surface modification with boronic acid led to the red shift in the absorption/PL spectra while the edge modification with borinic acid led to the blue shift because of the presence of boron atoms [93]. The surface modification with boronic acid abruptly increased the QY of GQDs to 29% in NIR with peak at 1024 nm [94]. The surface and edge modification of GQDs with boronic/borinic acid is represented in Figure 12. The change in band gap by these modifications is represented in Figure 13. In this figure, GQDs represents the pristine GQD, s-GQD represents boronic acid ($-BCO_2$) surface-modified GQD, s-B-GQD is boronic acid surface-modification with boron doped GQDs, e-GQD represents the edge-modified GQDs with borinic acid ($-BC_2O$), and e-B-GQD represents the borinic acid edge-modified GQDs with boron doping.

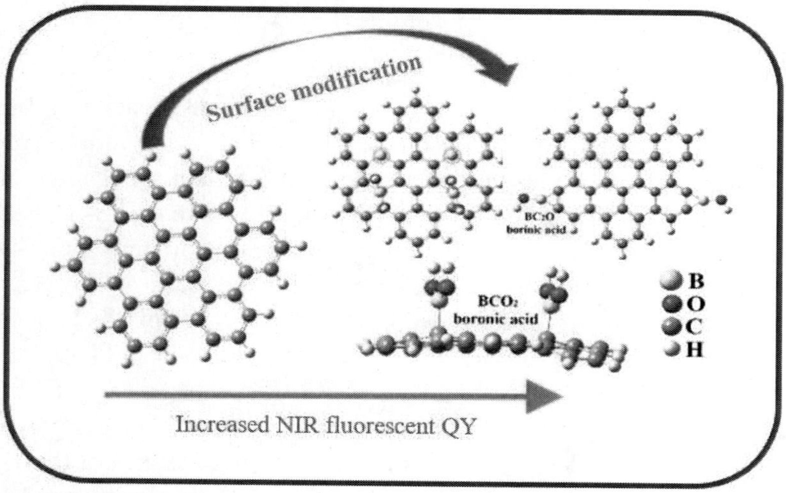

Figure 12. Surface modification of GQDs [94].

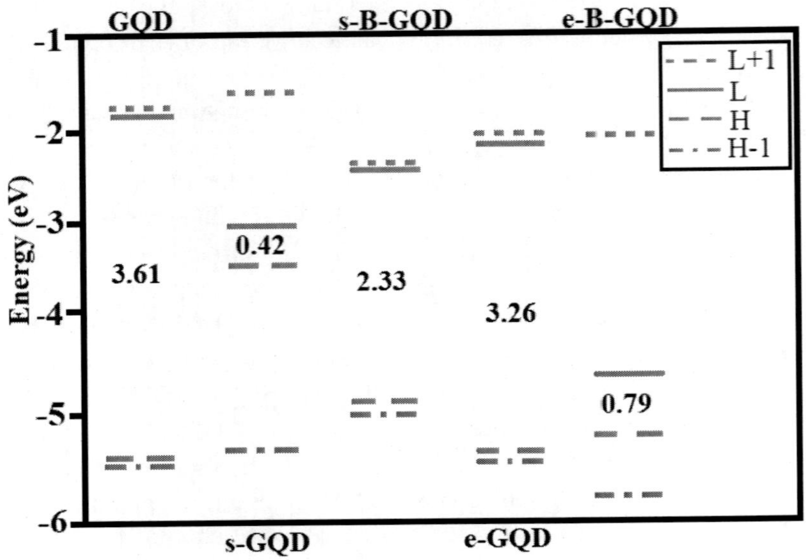

Figure 13. Band gap modification with surface and edge passivation of GQDs with boronic acid and borinic acid respectively [94].

3.5. Solvent Change

Like the other approaches, the solvent used for GQD dispersion is a valuable method for altering the characteristics of GQDs. Employing various solvents may result in the attributes that are required for a certain application. In part of the passivation on the surface of GQDs with the solvents, it is also shown that the PL of GQDs is sensitive to the aqueous media, especially water, acetone, dimethylformamide (DMF), and tetrahydrofuran (THF) etc. [61]. Furthermore, the protonation or deprotonation effects of pH fluctuation can change the PL intensity of GQDs [51, 61]. The variation in PL strength of GQDs with variations in the solvents or pH is mostly caused by their surroundings, which have an impact on the surface states of GQDs [90]. It has also been shown by the researchers that the dielectric constant of the solvent plays an essential role in modifying the optical and electronic properties of the GQDs. They studied the interaction of GQDs with heavy metals and it was found that as the solvent dielectric constant increased, the binding energies of cadmium (Cd) and mercury (Hg) cations were slightly raised while the Pb cation binding energy dropped [95]. Additionally, it has been demonstrated that the absorption spectra of GQDs exhibit a red shift in solvent compared to gas phase. And in solvent itself, absorption peaks are red-shifted in a solvent medium having large dielectric constant. The same has been studied by Zhao et al., they found that the absorption peaks of C_{132} GQDs are red-shifted when medium is changed from gas to toluene and from toluene to dichloromethane [96]. The type of solvent also has a remarkable effect on the fluorescence QY of GQDs. The QY values for the GQD suspensions in ethanol and water fluctuated between 15.2% to 21.6%. Additionally, GQDs dispersion in propylene glycol methyl ether acetate (PGMEA) showed the ultralow quantum yield (QY 2.5%) representing a significant fluorescence quenching. However, the addition of PEG coating greatly increased the amount of QY in PGMEA [97].

4. Applications of Graphene Quantum Dots

GQDs are potentially advantageous for a variety of applications due to their remarkable features and capacity to modify their attributes in accordance with a specific application by using any of the approaches mentioned above.

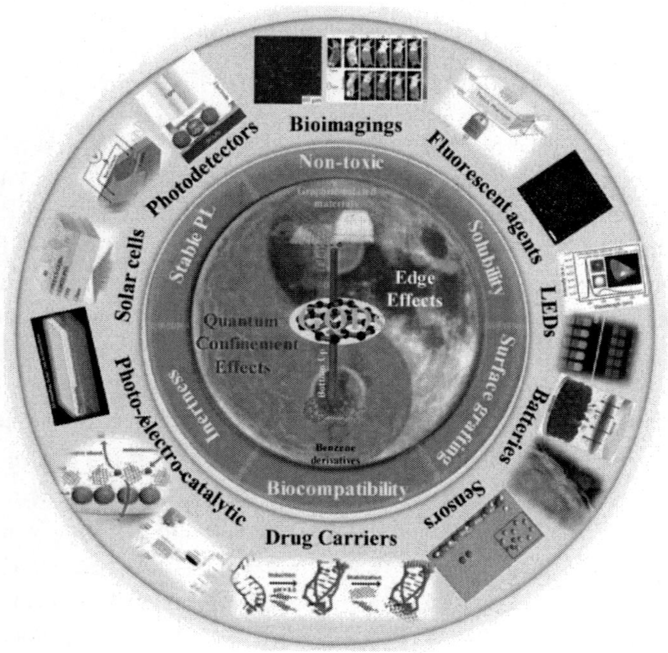

Figure 14. Various applications of GQDs [34].

To employ GQDs in applications like drug delivery, light-emitting diodes (LEDs), solar cells, etc., many researchers are committed to modifying their features. Additionally, the extraordinary qualities of GQDs, including their tiny size with QCE, excellent solvent dissolubility, photostability, non-toxicity, and furthermore, their capacity to migrate into living cells, make them promising components for cutting-edge utilization in the domains of sensing, imaging, theranostics, and many more [50]. Figure 14 displays a collection of GQD applications across multiple domains. A closer look at certain GQD applications has been included below.

4.1. Optoelectronic Devices: Solar Cells, Photodetectors, and Light Emitting Diodes

GQDs are used largely in the creation and development of the most practical optoelectronic devices, including outstanding durability solar cells, photodetectors (PDs), light-emitting diodes (LEDs), etc. GQDs are projected to be great for lightweight and portable optoelectronic devices towards the newest technology of smart innovations because of their exceptional optical qualities, incredibly thin layer, and tiny dimensions at the nanoscale level.

4.1.1. Solar Cells

Solar cells (SCs) offer a clean and renewable energy source since they can convert sunlight into electricity. Additionally, they serve as a vital energy source for societies with insufficient resources to produce electricity in the normal manner. Due to the features of GQDs, such as down-conversion, robust fluorescence, intense UV absorption, and ease of functionalization, many solar cells have been developed that use GQDs to improve the performance of the cells [98]. Compared to other conventional materials like silicon, perovskite, etc., GQDs garnered widespread interest in solar cell utilities because of their affordability, chemical stability, and lower toxicity [99]. Wu et al. used GQDs as an active layer of solar cells and demonstrated that the use of GQDs in hybrid solar cells increased the maximum power conversion efficiency (PCE) value by 40% [5]. This improvement was attributed to the improvement of the exciton dissociation and charge transport mechanism with the aid of GQDs serving as an electron acceptor and charge transport channel. A consistent film morphology of the active layer is obtained owing to the donor/acceptor's ability to mix uniformly as a consequence of the synergistic interaction between GQDs and (6,6)-phenyl-C_{61} butyric acid methyl ester (PCBM) used in the solar cell. Given their substantial optical absorbing capacity and almost optimal absorption in the visible and near-IR region, GQDs can be employed as sensitizers in solar cells by developing a unique solubilization technique for solution-processable GQDs [100]. The performance of solar cells can be enhanced by using GQDs as a photoanode. The use of GQDs as photo-anode layers in the SCs boosted the conversion efficiency to approximately 6.17% [101]. The early research and development of the function of GQDs in hybrid SCs would therefore enable comprehension of the functioning mechanism and future device performance enhancement.

4.1.2. Photodetectors

Photodetectors (PDs) which consist of highly photosensitive diodes with Schottky junctions or p-n junctions and normally work under external bias, transform the incident photons into electric current [102]. PDs are widely involved in optical networking, digital imaging, continuous surveillance, and other applications [34]. In the study by Zhang et al., high-performance PD in the deep UV region was developed using GQDs [103]. Meanwhile, the absorption of pristine GQDs is limited to this region therefore the operation of GQDs-based PDs is limited only to UV region. However, introduction of GQDs to some semiconducting materials or making their heterostructures (HSs) may enhance the performance of PDs. The HSs of GQDs composed of common semiconductor materials, like Si nano walls (NWs) [104], Si NPs [105], ZnO [106], etc., offered photodetection in the UV-NIR region. Additionally, photodetectors made of graphene/QD HSs performed significantly in the infrared spectrum [107]. The 2D HS of GQDs, such as GQDs along with MoS_2 [108], N-GQDs along with WSe_2 [109], GQDs along with graphene [110], revealed the high-performance wideband PDs in UV-vis-NIR. In comparison to pure Si NWs-based PD, Mihalache et al. stated 20 times greater photocurrent in GQDs along with Si NW core-shell HS-based Schottky PD with a notable augmentation of the external quantum efficiency. GQDs operate as a hole-blocking layer to improve charge dissociation in this system [104]. GQDs function as an active component for the amount of photocurrent present in these HSs. Nevertheless, the literature examines the use of both pristine and doped GQDs in the creation of PDs [111]. In order to improve the performance of the PD, Tetsuka et al. demonstrated an N-GQDs PD along with graphene-based PDs with a neutralizing boron nitride nanolayers [112]. Using an additional layer to improve PD performance is very costly for commercial applications.

4.1.3. Light Emitting Diodes

Another optoelectronic application of GQDs is light emitting diodes (LEDs), which employ GQDs as a charge-carrying layer or light translator. In order to boost radiative recombination, Song et al. manufactured LEDs based on GQDs, where GQD layers offer enhanced carrier transfer. This facilitated them to develop white light LEDs by using a current density [113]. Tang and coworkers revealed the construction of white light LEDs by covering commercial blue LEDs with extremely fluorescent GQDs. These GQDs changed the LEDs sharp blue emission into a broad spectrum that produced white emission [57]. As a result of the GQDs' coating, it should be noted that accumulation dramatically reduces their fluorescence intensity, which has a negative impact on the LED brightness. Researchers then employed an innovative approach ultraviolet (UV) photolithography which is a clever way to destroy such GQD aggregation and fabricated blue-light emitting reduced graphene oxide QD based LEDs with excellent solid-color emission [114]. The aggregation caused quenching (ACQ) can also be converted in aggregation induced emission (AIE) by properly functionalizing and reducing the size of GQDs as was done by Lee et al., who stated that the solid-state PLQYs of GQDs increased from 0.5 to 2.5% as their size is decreased from 5 to 1 nm, preventing ACQ. They functionalized the GQDs by two distinct rotor molecules and noted that all functionalized GQDs exhibit AIE behaviours with as much as 16.8% higher solid-state PLQYs. Systematic modification prevents inter- and intramolecular charge transfer, which increases the fluorescence rate of GQDs and the related PLQY, according to afterglow studies and theoretical calculations [115]. Thus, GQDs position themselves as viable material in organic and inorganic LED applications [34, 51].

4.2. Photocatalytic Applications

A chemical process that is promoted by a catalyst when it is exposed to light is known as a photocatalytic process [116]. The formation of charge carrier pairs by the incident light's absorption starts the photocatalytic reaction. Recently, this photocatalysis method has been utilized to detoxify the environment and produce sustainable energy [117]. TiO_2 has been utilized extensively for the UV-induced decomposition of dyes for environmental purification. TiO_2 has a significant bandgap, which reduces its photocatalytic effectiveness in the visible range [118]. Therefore, the bandgap of TiO_2 has been modulated in a variety of methods, including doping, functionalization, and HS production, to harvest the visible spectrum of sunlight [119]. A major focus in photocatalytic utilities has been devoted to the production of HSs of GQDs. Out of these, the development of HSs incorporating GQDs has attracted enormous interest in photocatalytic devices [120]. According to Gone et al., heterojunction of TiO_2/GQDs improves visible-light photocatalysis of methylene blue (MB) via the dissociation of charged particles [121]. They found that the HS had a 5.2 times greater degradation rate constant than bare TiO_2. One more shortcoming of semiconductor photocatalysts is fast rate of electron hole recombination [122]. GQDs overcomes these short comings of semiconductor photocatalysts. The naturally occurring neutral carbon in GQDs stabilizes chemical and physical characteristics and allows significant advancements in the manufacturing of standard photocatalysts. The following three factors can increase the photocatalytic effectiveness of GQDs combined with semiconductor materials (SCs): expanding active sites, broadening the range of light absorption, and speeding up charge transfer [123]. Doping in GQDs is also useful to make them viable candidate for using in photocatalytic applications. It has been reported in the study that the band gap of sulfur and nitrogen co-doped GQDs is about 2.54 eV [98]. These GQDs are generated by using a hydrothermal approach to produce green nanomaterials for the degradation of organic dyes under visual irradiation. A potential option for the breakdown of organic contaminants is provided by this unique photo catalyst [124].

4.3. Energy Storage Applications: Supercapacitors and Batteries

In order to address the ever-present energy conundrum, experts are focusing a lot of emphasis on renewable energy sources nowadays [125]. A further crucial problem is the storage of energy. In that perspective, lithium-ion batteries (LIB) and super capacitors are emphasised as prospective technologies for effective energy storage.

4.3.1. Supercapacitors

GQDs have piqued plenty of research interest in supercapacitors and batteries because of their distinctive structural characteristics, notably their large surface area [126]. Their electrochemical effectiveness illustrates the GQD abilities for energy storage applications. The GQD material used for electrodes exhibits an optimal electrical double-layer capacitance response with a considerable precise capacitance of 296.7 F g^{-1}, an acceptable energy density of 41.2 W h kg^{-1} at 1 A g^{-1}, a quick relaxation time, and outstanding cyclic stability [127]. Due to their nanometer size, numerous edge sites, and high surface area, GQD electrodes offer a desirable electric double-layer capacitor behaviour having surface capacitance ~192.9 F g^{-1} at 100 M v s^{-1}, which is twice larger when compared to graphene electrodes, according to the research by Zhang et al. [127]. Electroactive surface area of GQDs can be expanded when conducting polymers are introduced, which enhances the electrochemical behaviour and flow of current. In order to create solid-state supercapacitors, Wang et al. created polyaniline (PANI)/GQD-rGO/CFC (carbon fiber cloth) electrodes via spraying approach [75]. On the GQD-rGO-modified CFC surface, the various layer of PANI was afterwards coated and self-assembled. The multilayer of PANI/GQD-rGO were formed on the CFC surface by repeating this self-assembly process using spraying methods. In this approach, by using the H_2SO_4/PVA (polyvinyl alcohol) gel, which functions as both a binder and an electrolyte, a flexible and reliable, and powerful CFC capacitor was developed. At a current density of 0.1 mA cm^{-2}, this innovative capacitor had a highest capacitance strength of 82.9 mF cm^{-2} (1036 F g^{-1}). 97.7% of the charge was retained after 10,000 charging-discharging cycles, which demonstrated outstanding stability [128, 129].

4.3.2. Batteries

Because of its high efficiency, prolonged cycle life, light weight, high energy density, and durability over time, the LIB is one of the most useful ones for everyday portable devices [130]. The advancement of the electrode is highly desired since LIB performance is heavily dependent on its electrode qualities. Carbon-based compounds like graphite and graphene have long been thought of as LIB anodes ever since Akira Yoshino originally suggested using them as anodes in lithium-ion batteries (LIBs) in 1985 [131]. The combination of conductive carbon nanotubes (CNT) with exceptionally active GQDs in LIB anodes led to an impressive particular capacity (700 mAh/g at 100 mA/g after 100 cycles) and throughput performance [132]. GQDs with aforethought functional clusters provided substantial storage facilities for Li-ions, and the 3D CNT structures prevented the accumulation of GQDs. Anodes for LIBs made of hierarchical nanoflake-based CuO microspheres (H-CuO) covered with graphene quantum dots (GQDs) not only have reduced charge-transfer impedance and increased electrical permeability, but they also do not dissolve or clump together [133]. After 200 cycles at 0.2 Ag^{-1}, the bidirectional battery capacity of the GQDs/H-CuO hybrid anode approaches 609 mAh g^{-1} (pure H-CuO: 61 mAh g^{-1}). It also demonstrates permanent cycling stability, with an increased beginning coulombic effectiveness of 88.2% (pristine H-CuO: 75.2%) and a capacity maintenance rate of 79.4% after 1000 cycles (pristine H-CuO: 0.7%) at a substantial current density (2 A g^{-1}) [134].

Thus, strategic application of GQDs in supercapacitors and LIBs could possibly lead to the development of next-generation storage technologies.

4.4. Bio-Medical Applications: Bio-imaging, Bio-sensing, and Drug Delivery

In biological applications, ranging from bioimaging and injecting drugs in living cells, GQDs have received a lot of interest.

4.4.1. Bio-Imaging and Bio-Sensing

The development of GQDs is attracting interest in the disciplines of biosensing and bioimaging due to their outstanding photobleaching resistance, excellent water solubility, persistent photoluminescence (PL), minimal cytotoxicity, and good biocompatibility [135]. Long-wavelength fluorescent compounds are expected to provide better light penetration into the specimens and greater image contrast since biological specimens have a background of light scattering and auto-fluorescence in the short-wavelength region. The QY of GQDs is still extremely low, and in NIR, it is only 7%, while having in the visible region QY as high as of 98.5% [136, 73]. However, a number of changes to GQDs raised their QY in NIR. To address this need for bio-imaging, sulphur and selenium were doped onto the surface of GQDs, producing 26% fluorescent QY with respective fluorescence wavelengths of 850 nm and 800 nm [87]. Boronic acid surface modification of GQDs changes their fluorescence to NIR with fluorescence emission at roughly 1026 nm and fluorescence QY of 29%, making them particularly suitable for bioimaging applications [94]. In practice, doped functionalization GQDs with red emission are widely employed for bio-imaging applications as low energy excitation is advantageous for live cells [137, 138].

4.4.2. Drug Delivery

The internal biologic barriers make free-drug medications to living cells inactive and ineffective, whereas conjugated drug delivery may enter the cells and specifically reach the intended target. GQDs are now employed to deliver specific drugs to living cells. In addition to real-time drug release, Chen and colleagues disclosed nuclear-targeted drug delivery using GQDs coupled peptide transactivator of transcription (TAT) [139]. GQDs served as the drug carriers in the drug delivery mechanism, while TAT assisted in securing the medication to the cell's nucleus. Due to the drug blockage of Forster Resonant Energy Transfer (FRET) when it was released, the fluorescence signal of GQDs rose, and real-time monitoring was feasible based on a shift in the fluorescence intensity of GQDs. In tumors of the pancreas cells, hyaluronic acid-functionalized albumin nanoparticles that have been GQD-conjugated were successfully employed for drug transport, and green-fluorescent GQDs verified the existence of albumin nanoparticles were verified by green fluorescent GQDs [140]. As a result, GQDs have enormous promise for bioimaging and target-specific medication delivery.

5. Challenges

Although GQDs are promising candidate for various applications, still they are not used in many practical applications till now. There are some challenges in making their use in daily life applications, some of these are discussed here and indicated in Figure 15. First and foremost, their synthesis at large scale is completely challenging with the required properties for a precise application. The theoretically studied doping location is quite difficult to achieve experimentally to reach out the theoretically studied properties.

For bioimaging and bio-sensing applications, it will be very useful if the fluorescence is in NIR, but as discussed above experimentally achieved FL QY of GQDs in NIR is only 7%, theoretically it has been reached to 29% by surface modification of GQDs. Although it is quite high and may prove beneficial for such applications, but it has not been verified experimentally till now. Therefore, some work should be carried out in this direction to achieve such high fluorescent GQDs in NIR for bio-imaging applications.

Also, supercapacitors based on GQDs have seen some recent breakthroughs, still there are numerous obstacles to overcome. The incomprehensible size range of GQDs is one of the major problems. These issues make it difficult to comprehend the precise physiochemical characteristics of GQDs. Furthermore, its use in the electrochemical sector is restricted by its poor stability, lack of porosity, low conductivity, and asymmetrical redox sites. Thus, to improve its electrochemical performance, GQDs must be modified.

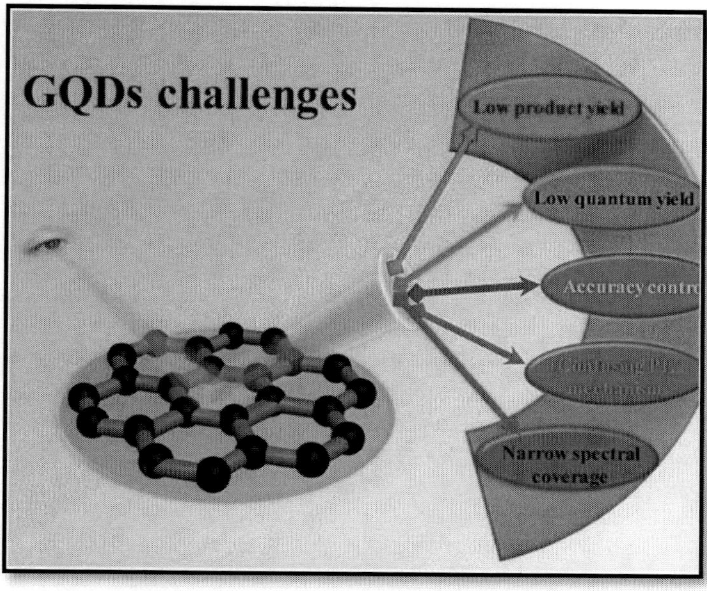

Figure 15. Various challenges of GQD [34].

In the realm of optoelectronics, GQDs are employed as a charge-transporting layer in the production of solar cells and other devices. Additionally, these devices often employ charge-transporting components that are readily available in the marketplace. The usage of pristine GQDs or hybrid architectures in optoelectronic gadgets avoiding any kind of industrial charge-carrying layers may present novel possibilities.

References

[1] Wang, X. W. et al. *Chem. Soc. Rev.* 45, 2239 (2016).
[2] Tepliakov, N. V. et al. *ACS Nano* 13, 10737–10744 (2019).
[3] Sk, M. A. *et al. J. Mater. Chem. C* 2, 6954 (2014).
[4] Jang, Y. et al. *Chem. Commun* 53, 1002-1024, (2017).
[5] Wu, W. et al. *ACS Omega* 4, 16159–16165 (2019).
[6] Yin, L. et al. *RSC Adv.* 9, 9301–9307 (2019).
[7] Ratre, P. et al. *ACS Omega* 7, 39586–39602 (2022).
[8] Reiss, P., Protiere, M. and Li. L. *Small* 5, 154–168 (2009).
[9] Nasir, S. et al. *Materials* 11, 1–24 (2018).
[10] Tian, Q. et al. *China Geol.* 5, 722–733 (2022).
[11] Wang, Y. and Hu, A. *J. Mater. Chem. C* 2, 6921–6939 (2014).

[12] Brodie, B. C. *Phil. Trans. Roy. Soc. Lond.* 149, 249–259 (1859).
[13] Haenni, V. K. and P. Z. *Anorg. Allg. Chem.* 105, 121–144 (1919).
[14] Bernal, J. D. *Proc. R. Soc. Lond. Ser. A Contain. Pap. Math. Phys.,* 106, 749–773 (1924).
[15] Wallace, P. R. *Phys. Rev* 71, 622–634 (1947).
[16] Vogt, F., Ruess, G. *Monatsh. Chem.* 78, 222–242 (1948).
[17] Offeman, R. E. and Hummers, W. S. *J. Am. Chem. Soc.* 80, 1339–1339 (1958).
[18] Boehm, H. P. et al. *Anorg. Allg. Chem.* 316, 119–127 (1962).
[19] DiVincenzo, D. and Mele, E. *Phys. Rev. B* 29, 1685–1694 (1984).
[20] Kroto, H. et al. *Nature* 318, 162–163 (1985).
[21] Mouras, S. et al. *Rev. Chim. Miner.* 24, 572–582 (1987).
[22] Novoselov, K. S. et al. *Science* 306, 666–669 (2004).
[23] Neto, A. H. C. et al. *Rev. Mod. Phys.* 81, 109–162 (2009).
[24] Lee, C. et al. *Science* 321, 385–388 (2008).
[25] Nair, R. R. et al. *Science* 320, 1308–1308 (2008).
[26] Ren, S., Rong, P. and Q. Yu, Q. *Ceram. Int.* 44, 11940–11955 (2018).
[27] Tiwari, S. K. et al. *J. Sci. Adv. Mater. Devices* 5, 10–29 (2020).
[28] Meric, I., Han, M. Y. and Young, A. F. *Nat. Nanotechnol* 3, 654–659 (2008).
[29] Ponomarenko, L. A. et al. *Science* 320, 356–358 (2008).
[30] Sharma V. and Jha P. K., Structural, Electronic and Vibrational Properties of Functionalized Graphene Quantum Dots Using Density Functional Theory. (2020).
[31] Rajender, G. and Giri, P. *J Mater. Chem. C* 4, 10852–10865 (2016).
[32] Zhu, S. et al. *Nano Today* 13, 10–14 (2017).
[33] Wang, S. et al. *Nanoscale* 8, 7449–7458 (2016).
[34] Tian, P. et al. *Mater. Today Chem.* 10, 221–258 (2018).
[35] Zheng, X. T. et al. *Small* 11, 1620–1636 (2015).
[36] Song, S. H. et al. *RSC Adv* 6, 97990–97994 (2016).
[37] Tretyakov, A. et al. *Phys. Chem. Chem. Phys.* 20, 14740–14752 (2018).
[38] Kim, S. et al. *ACS Nano* 6, 8203–8208 (2012).
[39] Dervishi, E. et al. *Nanoscale* 11, 16571–16581 (2019).
[40] Kim, J. et al. *ACS Omega* 3, 17789–17796 (2018).
[41] Yamijala, S. S., Bandyopadhyay, A. and Pati, S. K. *Chem. Phys. Lett.* 603, 28–32 (2014).
[42] Ali, M. et al. *Carbon N. Y.* 171, 493–506 (2021).
[43] Lai, J. et al. *Joule* 2, 76–93 (2018).
[44] Bera, D. et al. *Materials* 3, 2260–2345 (2010).
[45] Cassidy, J. and Zamkov, M. *J. Chem. Phys.* 152, 110902 (2020).
[46] Yan, X., Cui, X. and Li, L. S. *J. Am. Chem. Soc.* 17, 5944–5945 (2010).
[47] Arbiol, J. et al. *Mater. Today* 16, 213–219 (2013).
[48] Chukwuocha, E. O. et al. *World J. Condens. Matter Phys.* 02, 96–100 (2012).
[49] Berends, A. C. and Donega, D. M. C. *J. Phys. Chem. Lett.* 8, 4077–4090 (2017).
[50] Chung, S., Revia, R. A. and Zhang, M. *Adv. Mater.* 33, 1–26 (2021).
[51] Li, L. et al. *Nanoscale* 5, 4015–4039 (2013).
[52] Gan, Z., Xu, H. and Hao, Y. *Nanoscale* 8, 7794–7807 (2016).
[53] Chen, W. et al. *Nanotechnol. Rev.* 7, 157–185 (2018).

[54] Chhabra, V. A. et al. *RSC Adv.* 8, 11446–11453 (2018).
[55] Hai, X. et al. *J. Mater. Chem. B* 6, 3219–3234 (2018).
[56] Peng, J. et al. *Nano Lett.* 12, 844–849 (2012).
[57] Tang, L. et al. *ACS Nano* 6, 5102–5110 (2012).
[58] Tan, X. et al. *Chem. Commun.* 51, 2544–2546 (2015).
[59] Gan, Z. et al. *Adv. Opt. Mater.* 1, 554–558 (2013).
[60] Shen, J. et al. *Chem. Commun.* 47, 2580–2582 (2011).
[61] Zhu, S. et al. *Chem. Commun.* 47, 6858–6860 (2011).
[62] Sebastian, D. et al. *J. Phys. Chem. C* 126, 11182–11192 (2022).
[63] Fries, F. and Reineke, S. *Sci. Rep.* 9, 1–6 (2019).
[64] Yang, S. et al. *J. Mater. Chem. A* 2, 8660–8667 (2014).
[65] Xu, A. et al. *Synth. Met.* 244, 106–112 (2018).
[66] Kumar, Y. R., et al. *RSC Adv.* 10, 23861–23898 (2020).
[67] Rani, P., Dalal, R. and Srivastava, S. *Proc. 66th DAE Solid State Physics Symposium*, 562–563 (2021).
[68] Rani, P. et al. *Phys. Chem. Chem. Phys.* 18, 26232–26240 (2022).
[69] Liu, Z. et al. *Molecules* 26, 1–10 (2021).
[70] Eda, G. et al. *Adv. Mater.* 22, 505–509 (2010).
[71] Yan, Y. et al. *Adv. Mater.* 31, 1–22 (2019).
[72] Cai, J. et al. *Chem. Eng. J.* 435, 131833 (2022).
[73] Biswas, M. C. et al. *ACS Mater. Lett.* 3, 889–911 (2021).
[74] Zhu, C. et al. *J. Mater. Chem. C* 3, 8810–8816 (2015).
[75] Gong, X. et al. Synthesis of 3D nitrogen - doped graphene quantum dots and reduced graphene oxide composites by a hydrothermal method for supercapacitors anodes. *Ionics (Kiel).* (2023)
[76] Gorji, Z. E. et al. *Environ. Sci.* 126, 408–422 (2023).
[77] Qin, X. et al. *Nanoscale* 15, 3864–3871 (2023).
[78] Ganganboina, A. B., et al. *Biosens. Bioelectron.* 181, 113151 (2021).
[79] Chatterjee, M. et al. *Sci. Rep.* 12, 1–10 (2022).
[80] Wang, H. et al. *Nanoscale Horizons* 5, 573–579 (2020).
[81] Chalani, M., Daneshvari, V. and Hosseini, M. G. *Fullerenes Nanotub. Carbon Nanostructures* 30, 853–862 (2022).
[82] Huang, L. et al. *Chem. Eng. J.* 440, 135800 (2022).
[83] Yang, S. et al. *Chem. Mater.* 27, 2004–2011 (2015).
[84] Pan, Y. et al. *Mater. Chem. Phys.* 299, 127522 (2023).
[85] Li, X. et al. *J. Mater. Chem. C* 1, 7308–7313 (2013).
[86] Tang, L. et al. *J. Mater. Chem. C* 1, 4908–4915 (2013).
[87] Rani, P., Dalal, R. and Srivastava, S. *Phys. Scr.* 98, (2023).
[88] Gong, L. et al. *ACS Appl. Nano Mater.* 4, 8564–8571 (2021).
[89] Gil, H. M. et al. *iScience* 24, (2021).
[90] Zhu, S. et al. *Adv. Funct. Mater.* 22, 4732–4740 (2012).
[91] Jin, S. H. et al. *ACS Nano* 7 1239–1245 (2013).
[92] Yan, Y. et al. *ACS Nano* 12, 3523–3532 (2018).
[93] Feng, J. et al. *J. Mater. Chem. C* 7, 237–246 (2019).
[94] Rani, P., Dalal, R. and Srivastava, S. *Appl. Surf. Sci.* 609, 155379 (2023).

[95] Shtepliuk, I., Khranovskyy, V. and Yakimova, *Phys. Chem. Chem. Phys.* 19, 30445–30463 (2017).
[96] Zhao, M. et al. *ChemPhysChem* 15, 950–957 (2014).
[97] Yang, P. C. et al. *Nanomaterials* 11, (2021).
[98] Diao, S. et al. *Nano Energy* 31, 359–366 (2017).
[99] Bak, S., Kim, D. and Lee, H. *Curr. Appl. Phys.* 16, 1192–1201 (2016).
[100] Yan, X. et al. *Nano Lett.* 10, 1869–1873 (2010).
[101] Kim, J. et al. *Bull. Korean Chem. Soc.* 40, 56–61 (2019).
[102] Rogalski, A. et al. J. *Sensors* 23, 280–315 (2023).
[103] Zhang, Q. et al. *ACS nano*. 9, 1561–1570 (2015).
[104] Mihalache, I. et al. *ACS Appl. Mater. Interfaces* 9, 29234–29247 (2017).
[105] Kim, S. et al. *Sci. Rep.* 6, 1–8 (2016).
[106] Dhar, S., Majumder, T. and Mondal, S. P. *Mater. Res. Bull.* 95, 198–203 (2017).
[107] Zhang, M. K. et al. *Adv. Opt. Mater.* 10, (2022).
[108] Chen, C. et al. *Sci. Rep.* 5, 1–9 (2015).
[109] Nguyen, D. A. et al. *ACS Appl. Mater. Interfaces* 10, 10322–10329 (2018).
[110] Chiang, C. W. et al. *ACS Appl. Mater. Interfaces* 8, 466–471 (2016).
[111] Kim, C. O. et al. *Sci. Rep.* 4, 4–9 (2014).
[112] Tetsuka, H., Nagoya, A. and Tamura, S. I. *Nanoscale* 8, 19677–19683 (2016).
[113] Song, S. H. et al. *Adv. Opt. Mater.* 2, 1016–1023 (2014).
[114] J. Wang et al. *ACS Appl. Mater. Interfaces* 14, 48976–48985 (2022).
[115] Lee, S., Lee, J. and Jeon, S. *Sci. Adv.* 9, 1–13 (2023).
[116] Zhang, Y. et al. *Scanning* 2021, 3839235 (2021).
[117] Naraginti, S. et al. *Chemosphere* 336, 139210 (2023).
[118] Dao, T. B. T. et al. *Chemosphere* 280, 1–41 (2021).
[119] Kang, X. et al. *Catalysts* 9, 191 (2019).
[120] Niazi, Z. et al. *Photochem. Photobiol. Sci.* 20, 87–99 (2021).
[121] Rajender, G., Kumar, J. and Giri, P. K. *Appl. Catal. B Environ.* 224, 960–972 (2018).
[122] Zhu, D. and Zhou, Q. *Environ. Nanotechnology, Monit. Manag.* 12, 100255 (2019).
[123] Cheng, C. et al. *J. Hazard. Mater.* 424, 127721 (2022).
[124] Selvakumar, T. et al. *ACS Omega* 7, 12825–12834 (2022).
[125] Kumar, J. C. R. and Majid, M. A. *Energy. Sustain. Soc.* 10, 1–36 (2020).
[126] Ansari, S. A. *Nanomaterials* 12, (2022).
[127] Zhang, S. et al. *ACS Appl. Mater. Interfaces* 10, 12983–12991 (2018).
[128] Wang, S. et al. *ACS Appl. Energy Mater.* 2, 1077–1085 (2019).
[129] Roy, E. et al. S. *J. Energy Storage* 39, (2021).
[130] Liang, Y. et al. *InfoMat* 1, 6–32 (2019).
[131] Dong, C. et al. *EnergyChem* 2, 100045 (2020).
[132] Zhao, X. et al. *Nano Res.* 13, 1044–1052 (2020).
[133] Wang, Q. et al. *J. Alloys Compd.* 590, 424–427 (2014).
[134] Kim, J. et al. *Compos. Part B Eng.* 222, 109083 (2021).
[135] Kortel, M. et al. *Micromachines* 11, 866 (2020).
[136] Fan, Z. et al. *RSC Adv.* 5, 19773 (2015).
[137] Ge, J. et al. *Nat. Commun.* 5, 1–8 (2014).

[138] Qu, D. et al. *Adv. Opt. Mater.* 3, 360–367 (2015).
[139] Chen, H. et al. *Nanoscale* 7, 15477–15486 (2015).
[140] Nigam, P. et al. *J. Mater. Chem. B* 2, 3190–3195 (2014).

About the Editor

Dr. Aleksey Kuznetsov obtained his Ph.D. in Physical Chemistry from the Department of Chemistry and Biochemistry, Utah State University, USA. He graduated in 2003, after 3 years of doctorate studies, with a specialization in computational/theoretical chemistry. In 2019, after several postdoctoral and visiting professor positions in Germany (including an Alexander von Humboldt fellowship stay), the United States, and Brazil, he acquired a permanent faculty position at the Department of Chemistry, Universidad Técnica Federico Santa Maria, Chile. At this position, he has been involved in the computational design of various complexes of porphyrinoids, including core-modified porphyrinoids, with nanoparticles, fullerenes, and graphene, along with computational studies of transition metal complexes, organic compounds with pharmacological applications, metal–fullerene complexes, compounds with anticorrosive properties, etc.

Index

A

B

C

D